Cosmology
and the Cross

Chris Edge

Edge
Publishing

Cosmology and the Cross

COSMOLOGY AND THE CROSS

Copyright © 2013 Edge Publishing

EdgeFamilyPublishing.com

ISBN: 0615946240

ISBN-13: 978-0-615-94624-5

Chris Edge

TABLE OF CONTENTS

Introduction 5

Culture and Chaos..9

Faith or Superstition? 17

Science and Truth..30

Science vs. Humanities..34

Aspirations of the Heart..38

Messages of Hope..41

Too Good to be True? 45

Inescapable Faith..48

God or Nothing..51

Trinity...55

Creation 68

Full Disclosure..73

An Unexpected Arrival...75

A Cloud of Witnesses - The Body of Christ.....................................80

Evidence of Inspiration..91

Where else can we go? 105

The Cross...108

Cosmology and the Cross

Introduction

My favorite author once said that he wrote books that he himself would like to read. As I enter my gray-haired years, I now have two motivations for writing. One is to share thoughts and insights that I hope might be helpful to my children and (someday) grandchildren as they attempt to navigate the myriad of beliefs and attitudes of our age.

The second is to write a book that I myself should like to read when I need encouragement and when I forget the words of hope I once knew. When there is a season of clarity and vision, it is helpful to capture the insight one has and write it down. It is then much easier when times are tough to remember the vision by reading the written word and continue moving ahead.

When I was young, I never documented what I was doing at work unless I absolutely had to. Now I document profusely, because a year later it may be difficult to remember exactly what I was doing. This is a good thing of course for others who might need to pick up where I left off.

There are wonderful authors who write about matters of faith from the point of view of general logic and their vast knowledge of the humanities. I unfortunately do not have such knowledge. I wish I did. I have learned however, how to evaluate a physical system, break it down into the basic attributes of its components, and predict its behavior in the form of software. This has served me well in my career. If there is one thing that I have learned to obey over the years, it is to observe all the facts and data, acknowledge any implications regardless of my preconceived notions, and derive the best possible mathematical conclusions from that data.

In accomplishing the above, I always try to incorporate what other experts have said or proposed. If their insights are sufficient, my work is done. If their proposals are contradicted by the data, I must respectfully disagree and propose an alternative explanation that appears to agree with the data much more accurately. This of course is risky for a scientific career. The experts may know something that you do not know, or your understanding of the data may be inaccurate.

However, sometimes even in the scientific world there exists the human phenomenon of "denial", usually because some information contradicts a fundamental assumption that at the time seemed reasonable and obvious to a large number of people. When proposing a better solution, my hope would be not to contradict the original elegant assumption, but merely to broaden the assumption in a way that still retains a strong sense of simplicity.

In my non-work life, the same principle applies. I try to observe all the facts and information regarding this world in which we live and this thing we call "life," human life in particular. I apply the principles of individuals with "street cred" when it comes to living life. I then add my own flavor to the bits and pieces that perhaps were never directly addressed in the past, or which are unique to my particular experience.

Chris Edge

Who are the ones with "street cred?" My journey began with C.S. Lewis, because the tone and nature of his books connected with my heart, and I could tell that here was a man who understood about what life and love was about. Since then, I have come to realize that there have been many individuals who seemed to discover the same secret of life – they were chipper, joyful individuals regardless of circumstances, they were other-centered, they extracted limitless energy from a source that inspired them to work night and day serving others without regret.

There is no guarantee that (just because these individuals *seem* to be good sources of explanation for the meaning of life) in fact the explanation actually makes sense. There is no guarantee that it puts the pieces of life's puzzle together. It would be surprising, but not impossible.

Fortunately, for me at least, the story they tell is a very self-consistent description of reality. Despite the good and the bad, the wonders and the horrors that exist in this world, life has a purpose and meaning. There are many questions that remain unanswered or unanswerable, but overall the picture is a very good picture and does make sense. I have not found an alternative world-view that even remotely comes close to answering life's big questions.

Finally, when I began this book I intended the target audience to be those who were technically oriented, like myself, and who were debating the viability of Christian faith. Since this book is intended for me as a target audience as well, I realize that I also wish to add insights that have been helpful to me in the past. I will so indicate when I make an "aside" comment that will be of more interest perhaps to the believer than to the "not yet" believer.

I hope that you enjoy or appreciate this book, regardless of your point of view before, during, or after reading it. I can only promise that it is, if nothing else, a work of love, written from the heart as well as the mind. If you find that it strengthens your existing way of thinking, or that it opens your mind to new possibilities, I shall be very happy indeed.

Chris Edge

December 29, 2013

Chapter One

Culture and Chaos

No two people will ever agree, I suppose, regarding whether our existing world culture is "good" or "bad". There are those who firmly believe that our current freedoms of expression and behavior are liberating us to be fully creative and happy. There are those who believe that these same freedoms are turning us into a self-indulgent population headed straight for hell. There are many who are somewhere in between.

Wherever you lie on the above spectrum, I would venture to say that when it comes to topics like faith or morals, for the average person, our world today is very confusing. With a few taps on a keyboard or a touchscreen, I can red or hear worldviews of reality that are scientific, political, religious, humanistic, etc. with significant gaps between them, many leaning towards either "liberal" or "conservative." I would even go so far as to say that it is chaotic. There is a tendency therefore to assume that none of these approaches are true or false, they are merely in different categories.

I use the word chaos when I compare the divergence in the world on topics of faith and morals as compared with a discipline such as physics. When I compare the content of classes and textbooks, there is very close agreement between them on topics that have been around for a while – Newton's laws, relativity, electricity and magnetism, and quantum mechanics. In the last and newest category, there is some disagreement regarding how to look at things – for example, whether the position and momentum of particles is uncertain like a fuzzy cloud, or whether they are in fact precise but merely "hidden" from us. However, the properties as described by mathematical expressions are fairly well accepted.

As we enter the realm of recent observations, requiring particle colliders costing billions of dollars, there are multiple theories that attempt to explain how fundamental particles all fit together. For now, there are still many unanswered questions, and questions that are in the process of being answered. However, there is reasonable agreement and unity on topics that have had time to mature and develop. More importantly, there is general agreement on approach regarding the validation of theories.

In general, the approach begins with some basic assumptions and laws. From these basic assumptions and laws one is able to derive mathematical descriptions of all kinds of specific situations. If a wide variety of different situations seem to display good agreement between theory and observed data, the basic assumptions and laws are considered "valid" in so far as they go.

For example, Newton's laws are valid in so far as they go – i.e. for "typical" situations we may encounter where the velocity of objects being considered is much less than the velocity of light. Basic quantum mechanics is valid in so far as it goes – where the objects involved are atoms, molecules, protons, neutrons, and photons and they are NOT being excited with trillions of volts of electromagnetic energy. So the pattern

Chris Edge

seems to be that earlier descriptions of reality are derived from very accessible data and information, some of which may be accessible to all human beings – like the fact that an apple striking ones head will have much more impact when falling from twenty feet away vs. five feet away. More generalized descriptions of reality evolve from less accessible data and information, requiring more complex equipment to create or to observe and measure an event.

Likewise, when you consider the realm of faith and morals, there are certain topics that are closely linked to daily life, and are relevant and obvious to all human beings whether they live in a remote jungle or in the middle of a large city, whether they existed in Egypt in 2000 B.C. or are living today in 21st century America.

For example in the case of morals, the concept of possession and ownership seems to extend to the oldest historical records. Likewise, the wrongfulness of someone stealing an item that is the rightful possession of another person, of identifying such an action as "bad", and of constructing laws and consequences for such an action, appears to extend to the beginning of human history. This is not surprising since even a very young human being such as a two year old will often have grasped the concept of something being "mine" and will NOT be pleased if something regarded as "mine" is taken away.

In the faith category, I imagine that both primitive and modern human beings share a common sense of amazement and wonder when looking up at the night sky. Relating the development and growth of a small child once again to the development of historical humanity, the questions often arise "Who made...?" and "Why does...?" with regard to

11

natural sights (such as stars, trees, and ourselves) and natural phenomenon (such as water flowing downhill, the tug of earth's magnetic field on a compass, or the experience of color).

I think those of us in the scientific realm have a rightful sense that asking questions is a good thing, and that questions should have an answer. If the answer is not readily available, seeking the answer and sharing the results of this effort with others is considered a noble pursuit. If this were not so, receiving tenure for a faculty position at a university would not be so dependent on publications.

We have a reasonable expectation that a coherent story exists that explains the "how" and "what" regarding phenomena we see or measure with our equipment. I would have to say that physics does a good job explaining a very large percentage of these observations. Physics and science in general explain a lot of things very well – as far as they go. It would be presumptuous of physics to make statements about matters of faith, since by definition physics concerns itself with "natural" phenomena whereas faith concerns itself with that which we cannot see with our eyes. Faith likewise should defer to science on matters of strong tangible evidence such as the age of the universe.

Just as I have a reasonable expectation that a coherent story exists that explains the "how" and "what" of the universe in which we live, I have a reasonable expectation that a coherent story exists that explains the "why" of our existence and that explains the "big picture" regarding "what is this all about?" Since life to me is very important, I have a desire to know just how significant each individual is and why. I would also like to know whether this is all there is, or

Chris Edge

whether this thing we call "consciousness" and "life" - our "self" - will continue to exist, or whether it all ends at death. In particular, I would like to know whether this thing we call "love" is something significant and bigger than ourselves, or merely a chemical reaction or neurons firing in a particular way. A professed "scientific atheist" I believe would support the latter assumption.

I suspect that, in fact, there are few hard-core atheists. If you consider yourself an atheist I would like to make an observation and a challenge.

It would seem to me that a "scientific atheist" would reject anything that could not be proved by hard science. They would argue that it is fruitless to believe in concepts like God or the soul because they cannot be observed or measured physically. By contrast, concepts like gravity are also invisible, but the effects of gravity can be demonstrated repeatedly anytime we wish.

Words like "soul" or "spirit" may seem closely linked to words like "ghost". It is understandable that these seem like unscientific concepts. However, words like "consciousness", "person", "dignity", and "love" I think are less of a problem even for people with scientific minds. The reason might be that we all experience directly these concepts as human beings.

I would say that anyone who can appreciate and validate the existence of *these* words is affirming something that the hard sciences will never identify. Even though a person may exist in the context of a physical body, the idea of a person who loves and has dignity is something beyond just the physical molecules within which they happen to exist. I would say you

13

have to make a leap of faith beyond scientific data to acknowledge something called love and personhood. I think it is reasonable to say that the soul is simply that part of the person not limited by the physical molecules it happens to inhabit at this moment in time.

The proof of the existence of persons who are comprised of invisible souls and visible bodies are the physical actions we perform each day, the most important of which involve love. This is similar to inferring the existence of invisible gravity by dropping a rock. I can choose to believe we are merely complex organic robots. However, I don't know any parents, including skeptics, who regard their children this way. If we are all organic robots, then by definition all relationships I have are "I-It" relationships rather than "I-Thou" relationships as reflected on by Martin Buber. Regardless of religious persuasion, I affirm the reality and the duty of all of us to engage in "I-Thou" relationships.

In Christian circles, we are encouraged (rightly so) to "walk the talk." We are encouraged to live a good life of love and service to others in order to be consistent with our proclamations of faith.

I would say that if anyone claims to be truly a hard-core atheist who claims not to believe in love, dignity, etc. but who nonetheless adores their spouses, children, and grandchildren, I would have to say that they are not "walking the talk", and I am very glad that they are not!

So it is difficult to get away from the fact that most of us want to be considered as caring individuals, whether we are spiritual Christians or analytical skeptics. However, when I introduce words like "person", "good", "caring", etc. I am beginning to leave the world of scientific atheism behind and to enter the world of intangibles that I cannot measure with a detector yet are part of my daily existence.

My challenge therefore to all who feel obliged to rely on the sciences alone for their definition of truth and reality is simply this – consider the option of living an integrated life, where science is trusted for that which it is designed to do – describe and characterize observed tangible phenomena, and where faith is relied upon for that which is outside the realm of science, namely the uniquely human experience of love, personhood, creativity, and the arts. Be open to allowing faith to answer the questions science was never intended to answer – the "why" and the "meaning" of our existence.

Put all that together and you have the best of all worlds – the inspired pursuit of knowledge that is meaningful and has a purpose. Integrate your scientific skill and your childlike faith so that the awe and wonder that you experience when you dig and dig and dig in your particular field of study also inspires you to think outside the box, propose the non-obvious, and discover new fundamental truths that are vindicated by the ability to predict new phenomena *before* they are observed.

That is my observation and my challenge. In order to proceed, however, I think it might be helpful to tear down any barriers or concerns one may have regarding faith. There is an implied objection in the modern world that in order to believe one must stoop or lower oneself to cling pathetically to superstition. Living a life of faith cannot be and must not be living superstitiously or else it is merely living a comforting lie in order to feel good or to make it through the day. Living a life of faith must be a rational response to a Person who has revealed himself profusely throughout human history.

So now, what about faith itself? Is faith merely superstition, or is faith an appropriate way of responding to a reality that is currently too big for me to grasp or understand at a particular point in my development as a human being?

Chris Edge

Chapter Two

Faith or Superstition?

"Faith is the assurance of things unseen," is a good, simple definition of faith found in the Letter to the Hebrews in the New Testament. Never does Scripture say that faith is mindless belief in something illogical. Neither does it say that faith is merely acknowledging things that are obvious. I will present a definition of faith that is drawn from my lifetime experience as a scientist. Please regard this as a description of the concept in order to get away from the idea that somehow faith is non-rational.

A description of faith that works for me is to say that faith is the acceptance of inspired truth that appears to be validated by reality. There are other dimensions of Christian faith beside that – things that pertain to personal relationship such as child-like trust and confidence. However, from the purely intellectual side of things, faith requires a quantum leap, an opening of the mind to possibilities outside of the obvious.

I'd like to consider the key words in the definition I have suggested – the words "acceptance," "inspired," and "truth."

Acceptance – by acceptance I mean proceeding forward under the assumptions suggested by that which requires the leap of faith.

Inspired – by inspired I mean a *eureka* experience, an "ah-ha!" moment, some new awareness that occurs in our minds. It involves many pieces of the puzzle that suddenly come together and make sense, whereas before they did not. Inspired implies an explanation or description that is non-obvious, and possibly counter-intuitive, yet does seem to answer many questions and explain many observations.

To the scientist, inspiration is always assumed to come from the inspired individual. To the Christian, the inspired individual is given assistance from an outside source, although in many cases, in particular in the case of skeptics, this may not at all be obvious because it may seem to come from the depths of our consciousness or from the depths of our souls.

> *Years ago, as a young man, I began to have a glimmer of conversion. One night around midnight, I found myself having a conversation with someone from my bed. I suddenly realized that not only was I praying, I was also experiencing some form of quiet response. Once I became aware of this, I objected to the quiet voice with whom I was conversing "How do I know I'm not just talking to my common sense?" The voice responded immediately "You can call me Common Sense if you like for now – but be sure to capitalize it."*

Truth – is a description of or a statement about reality that is valid as indicated by strong evidence to back it up, in order to differentiate it from opinion. The connection between a truth and the evidence may not be obvious – hence the need for inspiration in order for the truth to be discovered. However, once asserted, the truth should be able to stand on its own merits. The more you try to test it or disprove it with new evidence, the truer it becomes. Sometimes it is helpful when wondering whether a truth is really true to ask oneself "What is the alternative?" and "What would be the implications if the opposite of this statement were true?"

For example, I can reject propositions such as "nothing travels faster than the speed of light" or that "electrons behave like a wave as well as like a particle", but such a rejection would be fairly ludicrous since I depend each day on a huge network of technologies developed by people using calculations based on these assumptions.

Whether I am someone who has a deep understanding of the meaning and implications of these statements, or whether I am someone who has no clue and does not have the time, energy, or mental talent required to understand these things, it is a very rational choice to accept these statements as true because there is a large integrated community of intelligent people who affirm these statements and a vast web of technologies developed using these assumptions.

I take the same stance in the area of faith and morals. There may be a few well-known role models who have served the poor their whole lives with joy, and claim that atheism was their inspiration. I doubt that there are many. However, I regard it as quite rational to pursue the same source of inspiration that motivated people like Mother Teresa of Calcutta to live their lives serving the poor.

By the way, when I consider role models throughout the centuries like Mother Teresa, I know that there will always be those who (in retrospect) believe they have discovered all kinds of issues and faults with that individual. They may find fault with the movements they began, thereby appearing to tarnish their inspiring story. For example, there have been recent studies that claim to have compiled all sorts of criticisms about Mother Teresa, such as the lack of use of narcotics and pain-killers by the Missionaries of Charity, or the lack of transparency of funds from donations.

Without getting bogged down in arguments, let me simply say this: I have never visited one of the locations where the Missionaries of Charity do their work. I have never met Mother Teresa. However, I have read numerous first hand accounts by individuals who became quite familiar with the work they do. There is an overwhelming reputation that grew over Mother Teresa's lifetime. You can find countless books containing these first hand accounts at any bookstore or library. I have also read many books containing quotes and reflections from Mother Teresa. Her attitude and actions towards the poor and suffering are quite clear.

The same is true with just about any inspiring hero I might mention. I am inspired by Martin Luther King from his life, his words, and his actions. I am inspired by Mohatma Gandhi and by Nelson Mandela in the same way. These men have a reputation of heroism that has been captured in countless articles, books, movies, and documentaries.

I have also heard criticisms about these men, regarding their personal lives, relationships with spouses, etc. I don't have the time, the energy, or the inclination to focus on these possibly valid issues. In today's information age, you can always find negative information about famous people. For me, as a non-expert in the biography of these inspiring individuals, it is sufficient to focus on the obvious – their words, their deeds, and the dominant reputation of their lives as captured in their recorded words and in the majority of the written and spoken reports of first hand witnesses.

Regarding Mother Teresa, we have a similar case in point. I know enough about her history, her mission, and her objectives from her words and from her reputation that I am confident that the aforementioned negative reports have missed the point. The primary objective of her work was never to be a hospital. Her primary objective was to offer a home and a loving family, herself and her sisters, to individuals who were dying alone on the streets like an animal. She transported them to a beautiful, peaceful environment, tended their wounds, and comforted them as best as possible in the same way their own families would have done if they had had a family.

Regarding the non-use of narcotics, it seems to me that a group of defenseless single women living in the heart of the poorest and therefore the most dangerous neighborhoods of a teeming city have already made a sufficient act of faith. It is a miracle to me that the Missionaries of Charity appear to have thrived quite well under such conditions. If you now add on top of their vulnerability the presence of expensive, addictive, and highly sought after drugs like morphine, etc., they indeed would have to become a hospital, with all the overhead of security guards, locks, etc.

Hospitals already exist. There isn't enough money or resources in India to pick up every dying person in the street and place them in the hospital. There are many poor individuals who have no family even to comfort them at home while they die. Those poor are the primary target audience of Mother Teresa and her sisters.

Regarding the use of donations, I have never seen a commercial on prime time TV where Mother Teresa was asking for money. I have received countless solicitations in the mail from nearly every charity on the planet (so it seems at times). However, I have never received a solicitation from the Missionaries of Charity. Things may have changed, but I know that during her lifetime, Mother Teresa never wanted to go "corporate" and organize fund raising events. Any donations were freely given by individuals who were inspired by her work. Perhaps there was a lack of being organized on her part. However, I have no qualms that the money was used well.

So the reason I have beaten this dead horse is because there exists a community of heroes and role models who share a common theme. They have been inspired and motivated to live a meaningful, selfless life, and in doing so have touched many lives in a positive way, resulting in a particular reputation worthy of honor. The common theme shared by these individuals is very simple – they felt called by Christ and they followed Him. I think it is very important to be inspired by their words, their deeds, and their reputation without being robbed of their example by those who sell books or articles or who obtain academic achievement by investing time in researching their faults or deficiencies.

I am not saying we should ignore the truth about anything or anyone. I am saying that we should weigh criticisms made about those who have said and done great things against the vast positive contribution they have made, without which their reputation and inspiring story never would have been told. Cynical people love to remove all heroes and inspiring examples. I am a realist about human beings and their weaknesses – however I will not allow others to erase the legitimate inspiration that I receive by observing the lives of heroic human beings.

What are the basic statements of faith that inspire such behavior? Are they feel-good mantras or are they proclamations of a wonderful new awareness?

Christian beliefs about the "big picture" of our lives on this tiny planet are an affirmation of the best possible scenario that my hopes and dreams could ever imagine, given what we all know that is good and bad about the world in which we live. The message is good news – too good to be true some would say, yet it speaks to the longing that has resided in many peoples hearts.

For example:

1) Is life meaningful? Yes. (Good news.)

2) Does God exist, is He good, and does He care for us? Yes. (Very good news!)

3) Are we each as human beings significant? Yes. In fact we each have the highest possible significance that could be conceived – we are each made in the image of God. We also each have a unique mission and a reason why we were created.

4) Does our life continue after death? Yes. (Whew!)

All the above good news also has a flip side – it implies we have dignity, choice, and responsibility. It means we our actions, thoughts, and attitudes have significance and therefore have good or bad consequences. We cannot escape that if it is true that we are made in the image of God and share the characteristic of personhood.

So the flip side relates to the "bad" that we all know is very real in this world. There are many incomprehensibly bad things that people do to one another. There are pretty amazingly bad or inappropriate things that we all know live within our attitudes and inner lives.

Sometimes these bad things get completely out of control and become full-fledged addictions or obsessions that we cannot seem to shake off, even though they may destroy our lives and the lives of those we love. Or perhaps we feel trapped to do things we know are bad because of external circumstances, such as turning to drugs or prostitution when homeless on the streets of New York, or becoming angry or cold because we feel trapped in a bad marriage, even though coldness is not our normal reaction to other people that we know.

However, this acknowledgement of the "bad" in this world and within us leads to this further incredible question and amazing response:

5) Does God care enough to do something about the bad things that can make life awful at times? Yes - by entering into the very heart of our existence.

The Christian statement of belief says that God has responded to our plight. Amazingly, He did this in a way that doesn't simply wipe out our independence and ability to choose, i.e. by turning us into benign robots. Instead, faith proclaims that He sent His Son to become one of us, to share in the joys as well as the miseries of humanity, and to offer

Himself as a suffering substitute in order to rescue us from ourselves.

Regardless of whether you believe in the existence of Satan or other sources of evil, the bottom line is that we make good or bad choices, and are often our own worst enemy. The good news is that we have a rescuer to deliver us from ourselves, our enslavement to self-destruction, and thereby giving us freedom to be who we truly are.

If I came up with all this on my own you would be right in assuming that I must be smoking something. If there were no one else in the world who had embraced these statements of faith and who had lived an amazing life as a result of it, the odds would certainly be against me. It would be as though I had made my own statements about relativity or quantum mechanics with no basis, no data, and with no former knowledge leading up them.

Relativity and quantum mechanics didn't just magically appear. They were the result of observations and seeking the answer to unanswered questions. There is always a legacy of knowledge, new observations, and new questions that lead up to new discoveries. The miraculous part is when the new non-obvious inspiration is given to someone that explains very well both the knowledge that preceded it as well as the new observations that require explanation.

Likewise, Christian faith didn't just magically appear, nor the questions that point towards it. I, Chris Edge, didn't just make this up. Religions, philosophers, and seekers throughout history have tried to make sense out of our life experience. The questions have been out there for a very long time.

Uniquely with the Judeo-Christian tradition, there is a significant historical legacy of God intervening with history, gradually revealing the answers to humankind. There is a 4000-year history of human beings who have heard God's voice and chosen the way of faith. There has been a progression of awareness regarding God's existence, His nature, His plan for humanity. This progression has led up to the coming of Christ, since whose arrival we have divided history itself into B.C. and A.D. Since Christ's coming, there has continued to be a progression of understanding, and a huge legacy of individuals who have validated in their lives that this faith is true, and have been willing to sacrifice all in order to live that truth.

> *Oddly, the story began in a tiny region of territory in the Middle East. Who could have guessed 4000 years ago that this small region would continue to play a key role in global politics today? I can't help sometimes but be amazed by all the attention a small desert region of the planet can generate.*

Just as statements regarding quantum mechanics are validated by a network of intelligent people and by the technologies that stem from them, likewise the statements of Christian belief are validated by the network of countless generations of good, wonderful, and holy people who have embraced the faith and answered the call.

One of the indications to me that this is truth is the enormous amount of pondering and investigation that this has produced. There are countless volumes of reflection, interpretation, speculation, and study on the Scriptures and on the Christian beliefs that arose out of the Christian community. Some may feel that this is irrelevant and that such disciplines as Christian theology are merely self-fulfilling academic endeavors that keep theologians employed. One may also argue that there are countless religions and philosophies that disagree with Christianity and have their own theologians and writers.

To the second argument let me clarify, as have others before me, that there is significant agreement and overlap between Christianity and other respected religions and philosophies. In many ways, Christianity does not disagree with the key beliefs of other beliefs and philosophies, but rather tends to fulfill and complete them. The more vehement arguments will be by those who disagree with all religions and who reject the basic concept of faith in general.

So to the first statement that the vast amount of study in the area of Christian theology and spirituality is irrelevant, I would say that one can draw a comparison between the vast amount of study on the meaning of Christianity with the vast amount of study on the writings of Shakespeare. As one who is ignorant of the academic study of literature, I could make a statement like "I reject that Shakespeare was a great writer of the English language." There may even exist papers and books that claim that.

However, the indication to me as a non-expert in literature that indeed Shakespeare was a great writer of the English language is the fact that there appears to be a vast community of knowledgeable people who believe he was a great writer. To the extent that there are those who have attempted to disagree, it is surprising that they have bothered to make the argument if indeed the works of Shakespeare were no better than pop novels.

So here is a helpful and humbling reflection for me as a scientist regarding truth. It is not possible for me as a scientist to put a thermometer on Hamlet and proclaim that indeed it is a great piece of literature. Science works that way, the humanities do not. By definition, the humanities focus on things that pertain to being human. As such, it requires a credible human or community of humans to define what is worthwhile and what is not, what is worthy of honor and of delving into, and what is not. The realm of science is the operation of the universe. The realm of the humanities is the creative work of human beings. The realm of theology is the nature of God and human beings together.

So when a large community of intelligent believers proclaim "Here is truth – dig into it," we should certainly sit up and pay attention. The testimony of these individuals does not prove that their subject matter is true, just as the testimony of experts in literature does not prove that Shakespeare was a great writer. However, the mere fact that so much energy has been expended delving into these areas implies that *something* is there. It is certainly rational to assume that Shakespeare must be a great writer, even if I have never read him. It is equally rational to accept that Christian belief is true, or has a lot of truth in it, and should not be merely disregarded as primitive religion or superstition.

I will never experience the reality that Shakespeare was a great writer unless I take a fair amount of time and effort to read him, appreciate him, and preferably listen humbly to experienced experts who can help me see much more rapidly things that are obvious to them, but may not yet be obvious to me. Once I have at least become an amateur reader of Shakespeare, I can perhaps discover some of my own insights. Once I begin to do that, I will probably begin to experience the thrill and amazement of reading Shakespeare that has clearly inspired generations.

In the same way, I will never experience the reality of Christian truths unless I spend at least a little time each day reading the sources that tell the story, listening to speakers who explain the story, conferring with those who are at least knowledgeable amateurs, and most of all seeking the person in prayer about whom the story and the proclamations are about.

Again, none of the above is intended to be "proof." It is merely an indication of why accepting this inspired truth is *quite* sane and rational, as opposed to mere infantile clutching to comforting superstition.

Chapter Three

Science and Truth

I find it interesting that the criteria of truth that we apply in the sciences seems different than the criteria we apply to faith and morals. In the sciences we do not take the approach that all opinions are equally valid even if they are diametrically opposed to one another. We do not say "if you wish to believe the earth is flat, that's OK" or "relativity is counter-intuitive, so I understand if you don't agree with it." Rather we try to form a consensus as a scientific community about our current best understanding of reality as described by physics, chemistry, etc. as confirmed by the evidence that we have at this moment. We attempt to capture these insights in the form of papers and conferences, and we try to summarize them in the form of textbooks in order to pass on this knowledge to the next generation.

A good scientist shouldn't mock an individual who refuses to accept basic truths. Neither should he capitulate or pretend that it's fine. Hopefully he or she will respectfully and patiently explain the scientific evidence relating to what is being rejected.

Likewise, when it comes to the question of faith and morals, we should not assume that the only valid response is that all beliefs or non-beliefs are equally true. We have intelligent brains that can ponder, weigh, and compare. We have feeling hearts that intuitively tell us "this is good – this is bad." One option is to have no point of view, to throw in our lot with no one, and to avoid the questions mentioned before. However, for each one of us the questions will come to us if we do not address them first. The question of life's meaning, and the existence of God, and are we loved by Him will confront us, like it or not, when we die, or when someone we love very much dies.

So when asking the question "Is it true?" about a theological worldview or anti-theological worldview, consider asking whether there exists a community of those who accept the truth that extends far back into time and which has documented theological evidence.

Scientific evidence isn't proof that a scientific concept is real. Scientific evidence simply gives strong support that a scientific concept is real. Experience shows that we do well to accept supported scientific explanations, because it allows us to move forward, predict new observations, and to create useful new technologies.

Theological evidence is similar. Miracles are not a proof – they are signs, indications, and confirmations that God is present and active. If there were only one or two such stories regarding signs that God had given as evidence of His presence, one could write them off as flukes or coincidences. But there are many such stories spread out over thousands of years and hundreds of generations. If this whole thing were merely a conspiracy or mass delusion, it would have died out long ago.

We cannot claim that God has not reached out to us. The miraculous stories in the Old Testament extend back 3500 years. In particular, the Exodus from Egypt by the Israelites is considered authentic history by many. Jesus' full time job when He wasn't proclaiming the Kingdom of God was performing miraculous healings. Since the time of Christ, there are countless stories of His followers who have healed the sick or have performed miracles of beauty and wonder.

All such miracles were performed in the context of having embraced Christian faith as true. Just as accepting scientific explanation opens the door to understanding deeper truths and to creating wonderful new technologies, accepting the inspired truth of Christian faith opens the door to a wonderful universe that has new meaning and adventure. Think of all the art and music that has sustained the test of time that was inspired by Christian faith. Think of the servants of the poor like Dorothy Day or Mother Teresa, or individuals like Fr. Maximillian Kolbe who offered his life for another at the concentration camps.

Today, we take for granted things like public hospitals, schools, orphanages for the common person and for the poor, not only for the rich elite. How common were such institutions anywhere in the world prior to the Christian Church?

If there were any other worldview that combined giving life meaning and hope with theological evidence spanning 3500 years, it would be worth considering. If there were a worldview that elevated our personal worth and dignity higher than Christianity, it would be worth considering. If there were an individual who lived joyously and happily all his or her life while serving the poor and making them feel

Chris Edge

loved and cared for while proclaiming that faith in God is unnecessary, I would respectfully pay attention.

I have not encountered an alternative worldview that points in the direction of living well, of living a meaningful life, of finding joy even during suffering. For me the evidence is overwhelming. Embracing the truth results in not only allowing the pieces of the puzzle to come together, embracing the truth opens the door to relationship with someone who is the source of all love and who has always known and loved me whether I knew it or not. Discovering this all-encompassing love is a source of courage and strength that allows me to live life in a way that is beyond my personal capabilities and limitations.

Why say no? By all means, be a great scientist or engineer. Think out of the box and make wonderful discoveries. But let your humanity also thrive. Let the unique person you are become solid and real. This means loving others generously, discovering your unique calling in life and living it well, and learning to feed your mind without starving your heart. For those of us who lean toward technology, I think it is helpful to reflect upon the complementarity of the sciences and the humanities.

Chapter Four

Science vs. Humanities

It is very tempting as a scientist to feel that the hard sciences own a monopoly on truth and knowledge. What more is there to know once I can predict exactly how to get to the moon and back using Newton's laws, or create a computer chip comprised of microscopic components that depend on solid-state theory? If I can't measure it, how can I claim to understand it? The humanities may seem superfluous. This is unfortunately aggravated by today's economy where the few good jobs that are out there often involve technology. Many are the concerned parents whose child has chosen to become an English major or art major in college, for fear their son or daughter may not be able to find a job, unless the major is used as a temporary path to law school or med school.

Despite economic realities, it would be foolish to think that the sciences own a corner on truth and knowledge. A very large segment of academic research and knowledge throughout the centuries have been dedicated to art, literature, music, history, etc. Most universities offer a good balance between science and the humanities. This reflects very well, I think, the complementarity of human nature and of the human brain itself. There is a part of our brain dedicated to math and geometry. There is a part of our brain dedicated to verbal and artistic skills. For many of us, one part of the brain may be stronger than the other. But preferably, we shouldn't simply allow the weaker part of us to die.

The best of all worlds is to allow our brains to be as whole and integrated as possible. Allow the math and logic on one side to impose discipline, while allowing the verbal and artistic side to experience wonder and thereby to think out of the box. Einstein expressed that he would have the intuitive insight in his head for quite a while for a new discovery. However, his math was weak, and it would take a lot of trial and error to convert his insight into elegant math such that it could now be shared with and explained to others more easily. As he once said, "I need a large trash container in my office for my mathematical mistakes – there are many of them."

So the sciences and humanities are like the left and right lungs, to borrow an image from John Paul II regarding the East and West Orthodox churches. The sciences are good in that they are often practical as well as able to evoke wonder. The humanities can be practical as well, for example when used as a preparation for law school. There are also various companies that recognize the value of a degree in humanities for their type of work. However, in general, the focus of the humanities are things pertaining to what is uniquely human – our literature, art, music, culture, and history. The primary motivation of the humanities is to focus on those things that bring meaning and insight to life, whether they have practical use or not.

There is no need for the sciences to belittle the humanities because they do not rely on the scientific method in developing knowledge. Likewise, the humanities do not need to be snobby and say that technology is only good for making bombs and pollution. The correct approach, I believe, is for both disciplines to see the interdependence between them. The sciences need the humanities in order to bring sanity and morality into how technology is used. The humanities need the sciences in so far as a scientific approach is helpful in organizing and confirming our understanding of humanity, particularly in hybrid disciplines such as anthropology and psychology.

In the same way, there is no need for antagonism between the scientist and the theologian. On the contrary, science can bring new insights and awareness to theology, while theology can highlight the fact that not only are the new advances in science merely "interesting" they are in fact wonderful and amazing, and further demonstrate the subtle mind of God.

My encouragement to myself and to all who are technical people is to allow our selves to be Renaissance people, who are comfortable moving between the worlds of science and of the humanities. If you love to write books as well write code, then by all means, write books. If you love to dance ballet as well as synthesize reactions, then dance away. Most of all, let's not use our devotion to research or to developing a new operating system as a substitute for being a good husband, wife, or parent. Furthermore, if we have been blessed with a good career because of our bent toward technology, let's find little ways to help others who have not been so fortunate.

> *Ten years ago while in high school, our daughter started hauling leftover bread, vegetables, dairy, and assorted other food items from a large supermarket to*

a nearby food shelf. As she moved on to college and post-graduate studies, it fell upon us to continue this weekly ritual. Once a good work like this has begun, it is much harder to say no, as compared to not having begun it in the first place. So we allowed the "positive inertia" to continue. It's a small gesture, like putting pennies in a piggy bank. But after 10 years, it's gratifying to realize that 500 minivans of food have fed those who weren't sure how to pay the rent and also feed their kids.

I further encourage us all to be spiritual people as well technically savvy people. I once heard a Benedictine monk point out that if you are at least seeking the meaning of life, you are a religious person. If you believe that other people are worthy of your love and respect then you are already religious, whether you know it or not. You already have one foot in the door of faith. There is no thermometer that I am aware of that can measure love or measure the value of a human person.

Chapter Five

Aspirations of the Heart

Whereas the mission of science is to understand how the universe ticks, the mission of the humanities is to develop our knowledge of things that make human life worthwhile. Animals are content to find food, breed their young, and protect against predators. Human beings are inherently relational, and inherently creative for the sake of creating. The reason why starving artists and musicians are willing to starve is because their passion for art and music exceeds their desire for mere comfort.

This all comes back to the longing in our hearts for something more than mere survival. Sometimes as scientists we expend a huge amount of time and energy pursuing new discoveries with various motives all working together. Some of our motives are knowledge for the sake of knowledge. However, I have observed the desire for success to play a key part. Success may be in the form of a Nobel Prize, a permanent position at a prestigious university, or the highest promotion possible within a corporation.

I was deeply moved by reading the biography of Paul Adrian Morris Dirac, who predicted the discovery of the positron and therefore anti-matter. What I had not realized till I read his story is how much loneliness and sadness he apparently suffered until the age of 28 when he received the Nobel Prize for his prediction of the positron which he inferred from his most elegant Dirac Equation. Soon after, Dirac married the sister of a fellow famous physicist, Wigner. Together, Dirac and his wife had a small family, and a good life. Although he had an honorary fellowship at Cambridge, he was never quite as productive in subsequent years as he had been during his youth. I couldn't help but feel that the honors he achieved paled in comparison with the simple joys of married life, in contrast to the lonely difficult years he knew before.

There are many possible honors and successes in life – Nobel Prizes, election to high office, promotion to CEO, etc. None of these things will matter when our time comes. The Nobel Prize won't hold my hand when I'm lying on my deathbed. Basically, love is hard to beat. Nurturing loving relationships is its own reward, though you will rarely see it in the headlines.

So my conclusion is that the longing in each of our hearts is a longing for love and for relationship. It is also a longing for meaning and significance. The promotions of this world can never measure up to the type of significance we are looking for. As a friend of mine has put it very well when encouraging his wife at home with the kids, "the work I'm doing today will be in the landfill in ten years." This is a valid perspective, particularly for those of us in technology.

There is good news here for the Christian. We said before that God has entered the world to share in our joys and sorrows. This means that everything we do, day by day, is important and significant. If my job is to love my wife, raise my kids, and go to work every day in order to meet their bodily needs, then that is the will of God for me, and nothing else could be more important. If I do all those things for the sake of serving and honoring God who created me and gives me life and health, then my life and work are significant. Everything else – promotions, prizes, and honors are merely icing on the cake.

This means that I can work hard and work steadily to do my research, build on my knowledge, and allow my life's work to be a monotonically increasing function. It is not necessary for me to burn out, or to discard my career and family in some midlife crisis. I can do all this without sacrificing my relationship to my spouse or to my children. This is all because my excitement and motivation for the work I do is primarily derived from knowing that this is what I am meant to do – I am following Christ in the unique way to which I have been called.

Chapter Six

Messages of Hope

We live in a difficult world. A good world, to be sure, but a difficult one. Some of us experience this more than others.

As I look back on the most prominent communications between God and humanity, I am struck by a particular characteristic – they are all messages of hope.

The world was particularly difficult for the people of Israel at the time of Moses. There was nothing easy or fun about the slavery to which they had been reduced. God speaking from the burning bush said, "I have heard the cries of my people Israel." The message was one of hope that God would not abandon His people to their misery.

An angel spoke to Mary regarding her Son, "He shall be called great. You shall name Him Jesus for he will save His people from their sins." Later, the angels declared to the shepherds, "Rejoice, for today I bring you good news, which shall be to all people. For on this day in the city of David, a savior is born who is Christ the Lord."

When He began his ministry, Christ's primary message and that of his disciples when they were sent out was "The Kingdom of God is at hand."

What strikes me about these messages is that in one way or another they are all confirming that God is here, He is very present in our lives, and intends to bring about His kingdom as opposed to the current situation, which is definitely not His kingdom.

What is the kingdom of God? It is perhaps easier for me to answer this question for myself by considering what is NOT the kingdom of God. The torture and punishment of Israeli slaves was definitely NOT the kingdom of God. When I see pain, suffering, and misery due to sin and neglect, I know this is NOT the kingdom of God. When I see angry marriages falling apart, and children caught in the middle, I do not see the kingdom of God. When I see leaders posturing for war, I don't see the kingdom of God.

These messages of hope occurred after long periods of apparent inactivity on the part of God. The Israelis had been slaves for quite a few years when finally Moses encountered the burning bush. The last prophets with their associated signs from God had been dead for hundreds of years before an angel finally appeared to Mary and to the shepherds.

As such, those messages for those people so long ago are extremely relevant for our people today. Like the Israelites suffering under the Egyptians and later under the Romans, it is tempting to say today "Where is your God?" We live in a time when it feels like miracles don't happen, probably never did happen, and superstitious belief has been supplanted by objective science.

In reality, for those of us who take the time to delve into recent advances in science and who take the time to delve into recent developments in theology and spirituality, the

Christian world has never been a more wonderful place. The advances in science tell us about the constituents of the matter of which we are composed. We see the Standard Model and String Theory being debated and developed as we speak. The existence of the Higgs boson appears to be confirmed. The universe continues to expand as a result of the Big Bang, the existence prior to which science cannot yet really speak, and the acceleration of the galaxies seems to indicate dark matter or dark energy, which is still not understood. So we still have as many questions as answers.

All these amazing discoveries that are still in progress feed into the conviction on the part of theologians that the universe is not less mysterious for us than it was for our ancestors. Rather, the universe is more mysterious and wonderful than they could ever have imagined. Whether I am pondering the movement of galaxies or the mitochondria in the living cells of my child, there is no lack of material to feed my sense of wonder about life and about this universe. It is particularly amazing when I consider that there is no *a priori* reason why the universe should exist at all, that we should have well-defined physical laws, or that matter and structure should exist.

As James Taylor once wrote in his lyrics to "The Secret of Life",

> *Einstein said he could never understand it all*
> *Planets a-whirling through space*
> *The smile upon your face*
> *Welcome to the human race*

It is enough for me to experience the daily miracle of existence and the complexity of life as I see it in myself and in those whom I love. Even if I accept the conclusions of biochemistry that the development of life from amino acids has been a logical development within the time frame of a billion years, I still regard it as a happy accident. I also seriously doubt, although I could be wrong, that any computer model could be initialized with a description of amino acids a billion years ago and wind up predicting the creation of Hamlet or an iPhone without a little outside help.

So the messages of hope still remain relevant today in our skeptical, technically savvy world. To ignore or reject these messages leaves us alone with our technology and with our current scientific explanations that tell us something about how things work, but cannot tell us why we should care. The messages of hope tell us that God is still here, still just as active, right here in midst of our crazy technology that leapfrogs every ten years instead of every century. I love the scientific work that I get to do, but I will never pretend that it is a replacement for the ongoing word of hope humanity has received these last 4000 years.

Chapter Seven

Too Good to be True?

I would like to begin with a clarification, for myself, for those who are skeptical, and for those who possess a strong faith. The clarification has to do with attitude. What is God's attitude towards those who doubt, who struggle with or outright cannot believe the best of all scenarios that I listed in the previous chapter?

I am on retreat as I write this particular section. The image suggested for reflection on the last day has been the appearance of Jesus in the upper room to the disciples. In His first appearance, all eleven disciples were there except one, Thomas. The ten who were present were obviously overjoyed. Jesus whom they had followed for several years had died and now had miraculously returned. Despite the hints He had made before His crucifixion, the possibility that He could be executed and then return seemed out of the question. It was both too terrible and too good to be true.

Later, Thomas categorically could not accept the eyewitness account of his comrades. Despite being outnumbered ten to one, the skeptic in him was too strong. Note however that he did not say, "I will never believe no matter what." Rather he said, "I will not believe unless I see the mark of the nails in his hands, put my finger into the nail marks, and my hand into His side." Thomas wanted to be sure and expressed honestly what he needed in terms of data in order to be sure.

A week later, Jesus appears to the disciples again. This time, Thomas was there. Here is where I would like all of us, believers and doubters alike, to observe what Jesus did *not* say. He did *not* say, "Thomas, leave the room. Your doubt has disqualified you." Rather He said, "Put your finger here and see my hands, and bring your hand and put it into my side, and do not be unbelieving, but believe."

So I would like to point out that the disciples had no less difficulty with faith than we do in this high tech, skeptical world in which we live. They were no more prone to superstition or believing in magic shows than we are. Consider what happened after Christ appeared to Mary Magdalene and she immediately told Peter and the other apostles what she saw. Scripture makes it quite clear that the disciples couldn't believe her story because it seemed "like nonsense." It seemed too good to be true. Keep in mind that these guys had seen daily miracles performed by Jesus these last three years. They had even seen Him raise the dead multiple times, the last time being Lazarus who had been dead for three days and was already in a tomb.

Christ is patient with skeptics. Jesus wasn't resigned to the unbelief of His friends, but neither did He write them off. Instead, He challenged them to put away their skepticism and believe. God had done a great thing, and had presented them with a great gift. Jesus would not be satisfied until they had recognized and accepted this gift, because He knew it would become the core meaning of their lives and of the lives of many others.

I think many of us would like to believe that God exists and that He cares for us. We would like to believe that Jesus came to make whole our broken humanity and to empower us to live vibrant lives as children of God. If what Christians believe is true, God is fully aware of our struggles with faith and our skepticism. He was obviously fully aware of this with the disciples.

Even if you don't know how to pray, or what prayer is, I suggest simply admitting honestly like Thomas, "I cannot believe unless..." The "unless" part will be unique to every person reading this book. Perhaps you need some special signal between yourself and God. Perhaps you need freedom from a hopeless addiction. I know that if Christ was patient with the disciples, He will be patient with me and with you. Years ago, my prayer was, "If this is true, let me know it is true."

For me, the sign was simply certitude deep within, "This has to be true – it cannot be otherwise." Whatever it takes to make it happen for you, my prayer is that you receive it. The answer may be as simple as a peaceful, confident certitude or awareness that God is real, He is there, that He knows you and that He loves you.

Chapter Eight

Inescapable Faith

As scientists, we like to believe that we live by fact and data, nothing else. Faith does not come into the picture.

In reality, if we did not live by faith, we would not get very far as scientists. When I was a young physicist, I probably did not understand a majority of what I learned. I got by with passing exams. I could memorize certain equations, manipulate them, and learn to obtain the answers I needed. But there were many things I did not understand. Doubting whether they were correct or true did not enter my mind. There was no doubt that the deficiency was on my side, not on the side of the body of knowledge I was confronted with learning.

Over the years, two things have emerged. One is that the concepts I learned about in school had broad applicability in the real world of technology. I found that as I solved technical challenges using things I had learned, I finally "got it." Many concepts sank in and made really good sense to me. This came home to me one day driving home with my kids on a road trip. I had never had an opportunity to take a class in general relativity. The subject came up regarding black holes. So I found myself attacking the question on the back of a napkin of how dense must a neutron star become in order to become a black hole.

This would have been a daunting question to me in school. But over the years I had been confronted with many challenging problems that had to be solved from basic principles. So I began with the math I did know – special relativity which requires not much more than the Pythagorean theorem, extended it to include acceleration, associated that acceleration with gravity, and voila, before long I had an estimate for the density of a neutron star becoming a black hole. It felt pretty good!

So the broad applicability that I learned over the years in order to do my job gradually allowed me to understand many mathematical concepts in physics that I accepted but had no clue what they meant. Furthermore, as I have understood some of these concepts, I have realized that if someone had explained them to me a bit differently, I might have understood them the first time. Unfortunately, finding really good physics teachers can be difficult – it is a difficult subject to understand, and not many can explain it well in plain English. I often hear that people hated physics in high school. Now and then I hear someone say they loved it – because they had a really good teacher.

I am very grateful that I didn't give up and reject my subject of choice due to lack of understanding or due to teachers who had difficulty explaining. I hate to admit it, but part of the draw for me was just looking at physics equations and thinking that somehow they were "cool."

So I think we really can't escape faith. It is merely a question of where to put our faith, of where we sense that truth exists, as evidenced by the way people live and how they speak. This wasn't a difficult question for me to answer years ago. My two favorite authors by far - C.S. Lewis and J.R.R. Tolkien, were both profoundly Christian. From there I met individuals one after another in whom I saw a light, a joy, a love, and a kindness that I could not easily find anywhere else. I placed my faith in the messages of hope that they in turn had placed their trust, and never regretted it. There was much I did not understand – but I have found the truths from those messages to be broadly applicable in my life, and as I have applied them, they too have gradually "sunk in" and made very good sense to me.

Chapter Nine

God or Nothing

It's funny that sometimes in the midst of beautiful albeit (perhaps) sentimental songs one can find nuggets of profound wisdom. An example that comes to mind is "Something Good" from Oscar and Hammerstein's "Sound of Music":

> *Nothing comes from nothing*
> *Nothing ever could*

I look at the air around me and see nothing. I look at the sky and trees around me, and see only sky and trees, nothing more. I don't see God. I see no mysterious invisible hand at work, carving, shaping, and creating. I see only what I see.

The universe exists. I exist. Other people exist. Beyond that, what can I possibly say without being speculative, superstitious, or wishful in my thinking? The universe is ticking away, functioning like a giant jigsaw puzzle comprised of a myriad of physical laws that result in the world we see. There is a comfortable reliability and predictability about this universe and its properties. I appreciate this every day in my work because whenever things don't make sense, when code simply doesn't run correctly or the data simply doesn't agree with an expected result, I can be tempted to think that it is hopeless. The software will never work, or I will never obtain results that make sense.

As I say, there is comfort in our predictable universe. Because I can always say to myself reassuringly, "There *has* to be a reason for this apparent illogical malfunction. There is *always* a logical explanation. Be patient, and you will figure it out. When you do, it will make sense, and things will work properly."

Our universe has structure and form and as such is comprehensible. Even more surprisingly, we exist to comprehend it. This is by no means a given. Even if I look at all the non-human life that exists on this planet, there is no *a priori* assumption that a form of life will arise that has the luxury of pondering this universe and speculating about how it works.

I am often reminded of science fiction stories about planets comprised of androids whose creators have all died. They are tempted to assume that they have always existed, and that it is merely fanciful wishful thinking that flesh and blood originals might have been their creators. Or stories about multiple generations living on a space vessel headed for a far away star system who have forgotten their original mission. For them, the space vessel "simply is" and had no original creator.

I think of a person sitting on a rock for a billion years, waiting expectantly for an iPad or PC - complete with AC power, monitor, and keyboard - to arise out of the ground. Or the unlucky physicist sitting at the bar next to an unclaimed martini that he requested, hoping that the random molecules in the air will coincidentally coalesce into a beautiful woman who will happily accept his offer of a drink (the odds after all are at least 1 in 10^{30}!) When the bartender asks whether it was also possible that such a woman could simply walk through the door and sit down next to him, he morosely responds, "Yeah, like what are the odds of *that* happening?"

What I appreciate about all the above scenarios is that they highlight for me the miracle of creation, including our own. There is no loss in scientific integrity by saying that the scientists and the theologians are both right. Our physical existence is the result of scientific laws and phenomena. Our physical existence is part of an amazing process quietly guided and ushered by the hand of God. Both statements are true. It is no different than saying that our bodies and brains operate under physical laws, electro-chemical reactions, biological processes, all of which are scientific phenomena, but that somehow in the midst of these complex events a human person arises. Despite the valuable insights of science, we all operate as a society under the assumption that a human person is of infinite value and should be respected and protected. Reduce a person to mere chemicals and the value ceases to exist.

We cannot see a person. We can only see a person's body when it moves and speaks. Externally, when someone dies, the only difference we see is that the body no longer moves, and that which animated it is now gone. So I affirm the reality of persons even though I cannot really see them. I affirm that personhood, love, beauty, truth are invisible realities that are just as real and perhaps more important than much that we can see and touch. I therefore affirm that unlike the skeptical androids it is blatantly obvious to me that a Person is responsible for this amazing universe and that only a Person could have created me and the persons around me. Nothing comes from nothing. The options are God or nothing. I choose God.

Chapter Ten

Trinity

The Trinity is a unique concept within human history. There are a variety of monotheistic religions, many polytheistic religions, and various forms of mysticism that may refer to some type of life-force, spirit, etc. that may or may not be equated with God. Christianity is certainly unique in proposing a God who is one (i.e. in agreement with monotheism) but who is comprised of three persons. This latter refinement may sound quite like polytheism.

I believe that the critical distinction between three persons and three gods are found in Christ's statements like "We and the Father are one" and "we will send the Holy Spirit". The "we" who are not "two" but "one" will send the "Holy Spirit" referred to earlier in the Gospels as "filling" Jesus.

An old Dominican that I knew years ago once told me that the Trinity was a "gratuitous revelation". Whereas the Incarnation and our sanctification by Christ is an essential revelation that directly impacts our lives and decisions, God did not necessarily *have* to reveal to us that His essential nature is three persons in one God. His primary motivation for this disclosure, one may assume, is love. Love, if one believes in love, has the characteristic of wanting to know and understand the beloved, and to disclose and reveal ones essential self *to* the beloved.

Having received this supposed revelation, does it seem reasonable and does it make sense? Does it explain certain characteristics of God that we believe or would like to believe are true?

Even someone who cannot accept Christianity, but shares many of its values, may wish to believe in a loving God who cares. The most effective program (if not the *only* effective program) for becoming free of addiction *requires* acknowledgement that "I can't do it alone" and must accept help from "a higher Power" which clearly means some form of benevolent God, with no specific religion mentioned.

If God is less than a Person, then there is no possibility of relationship and no possibility of love. An object like the sun or a rock cannot love. A force like the wind or like gravity cannot love. The cosmos or the universe, with the exception of individual persons like us, cannot love.

Since many (if not most) good people would agree that persons, love, and relationship are the highest values one can have, would a God incapable of love and incapable of relating to others be in any reasonable manner something we would want to call God? Does it seem reasonable that we imperfect humans should be superior to this God, since we can love and relate to others, and He cannot?

To me the choice is either to embrace a personal God, or to believe in no God at all.

For those who at least acknowledge that belief in the existence of God is reasonable, the next question is why three in one? What is the alternative to three Persons? One could argue that classic monotheism is reasonable, which would say that God does indeed relate to us, and that he is therefore (at least) a Person.

Is he more than just a single Person? Here we come to the statement that many non-Christians might find very appealing and very comforting: God is Love. To say that God is Love seems very true to many people who have some form of belief. It is a very encouraging statement, especially if ones religious tradition or experience of religion was very negative, focusing too much on God's justice or punishment.

What is fascinating to me is that the plausible statement "God is Love" is in perfect harmony with the "gratuitous" revelation that God is three Persons in one. It is difficult to "be" Love if Love is only one person loving himself or herself. The Trinity makes the statement "God is Love" a powerful reality, not just a flowery phrase. Three persons with distinct identities can indeed love one another. The idea that somehow Love entails becoming "one" with the other while remaining distinct seems very compatible with the highest form of what Love is all about.

But why three? One might grant that *perhaps* the idea of two Persons in an infinitely unified relationship constituting one being makes sense. One might concede that the idea of more than just one Person is consistent with statements like "God is Love." Why not simply remain at two?

Since we are speaking here about persons and relationship, I can only draw from the data of personal human experience and the data of the unique characteristic of life itself. There are two concrete examples that I can think of that would support three persons rather than merely two.

One of the unique characteristics of life is that 1+1 does not necessarily equal 2. You and I are persons and are composed of billions of individual cells. I believe in the unique reality and the identity of the person known as Chris Edge, and presumably you feel the same way about your unique identity and existence. Most (if not all) of the actual physical cells that originally composed my body when I was born have been replaced. My self, my person, has remained intact. I do not regard myself as having been destroyed and replaced with a clone.

I believe in a single reality known as Chris Edge that has changed and evolved over the years, yet retains that unique identity of personhood. I therefore affirm that I am more than merely the sum of the billions of cells that comprise my physical body at any one point in time. I affirm that belief regarding you and regarding every individual on the planet. When it comes to life, especially human life and personhood, the sum of the parts may result in a person that does not exist with the same parts existing separately.

A second example of the whole being greater than the sum of the parts is the human experience of love and relationship. When two people love each other, as friends, family, or as lovers, it is not uncommon to sense that something bigger than merely the "we" or the "us" exists. For a family, there is a sense of identity of the family unit that goes beyond any one individual. For friends, there may be a special bond of kinship, life, creativity, or purpose that only occurs when the friends are together.

For lovers, there is a presence of love and magic that may seem like its own reality. Testimony to this can be found throughout literature, especially poetry, which can be a high form of recorded human experience. How many romantic poems throughout the history of every language and culture speak of Love, or Destiny, or Fate in association with the beloved? This Love, Destiny, or Fate is not associated with the individual lovers by themselves as individuals but rather in relationship with one another, and is associated with the goodness and value of that particular relationship.

Our human experience and the characteristic of life itself strongly supports the idea of a separate, good reality that exists as a result of the uniting of individual living cells as well as the uniting of individual persons. The claim of the revelation of a third reality and indeed a third distinct person arising from the loving, unified relationship of two persons seems to me to be not only quite reasonable, but also quite a *eureka* discovery.

When I reflect on the Trinity in the above manner, I experience a *eureka* moment similar to reflecting on the proposals of relativity or quantum mechanics. For me, at least, relativity or quantum mechanics are not nearly as self-evident or obvious as Newton's laws of motion. The idea that things move faster if I push on them harder and that objects speed up the longer they fall doesn't require a huge stretch of my faith. Newton's laws are merely a clarification and quantification of patterns of how things behave around me on a daily basis.

Relativity and quantum mechanics on the other hand explain things that I may experience but don't really think about, like why do I see colors in a rainbow or why does matter exist and remain intact. They also explain realities that require specialized equipment to observe, such as glowing light from burning materials that exhibit only *certain* colors of the rainbow when passed through a prism, or the unexpected bending of starlight near the sun during a solar eclipse.

The fact that these theories are not completely intuitive is evidenced by the time it took to make sense out of these phenomena and the time it took for the scientific community to agree with some of their basic assumptions. It is evidenced in how long it took to capture these descriptions in the form of concrete equations that can be compared with tangible data and used to make predictions.

The fact that so many experiments can be explained with the equations we do have seems to confirm that the proposers of these theories are on to something, and that these non-obvious theories are valid. Likewise, so many practical technologies and products are a direct result of these esoteric theories and equations.

To the extent that these theories and equations are explained and defined, they provide a *eureka* for those who have discovered them and for those who have taken the time to learn and understand what they are all about.

In the same way, it should come as no great surprise that the explanation that God is Trinity might be non-obvious and yet make excellent sense when we take the time to think about it. Likewise, although God could have handed a detailed description of His nature on a silver platter, it is not unreasonable that He should give us enough information about His Trinitarian nature in the recorded stories and words of Christ in the Gospels, but leave it up to us to make sense out of what was seen and heard.

God did not hand Sir Isaac Newton a tablet with the

equations and assumptions of Newton's laws. He allowed Sir Isaac to experience gravity and motion directly himself, and then gave him a good brain to create the names, the assumptions, and the written equations that we know as Newton's laws on his own, with a bit of help that we call inspiration.

In the same way, Christians experienced the words and actions of Christ directly. As time went on, for the purposes of communication, understanding, and the passing on of knowledge, Christians with the capacity to think about God (the way that Newton thought about the universe) created the names, assumptions, clarifications, and explanations regarding the Trinity in light of the recorded experience of the words and deeds of Christ.

One can object that the word Trinity came much later, and therefore was a fabrication from Christians wanting to create their own little theological system. However, just because a *eureka* comes later in time does not make it any less valid, as we have seen with relativity and quantum mechanics. The valid question to ask is not "when did the eureka occur?" nor is it "was the proposed theory obvious?" but rather "does the idea make excellent sense if you really take the time to think about it?" and "does the idea fit together very well with the data or information we currently have?"

I have already indicated that for me, at least, the Trinity supports and in no way contradicts basic statements like "God is Love", as well as the reality of the whole being greater than the sum of the parts.

I leave it as an exercise to the reader to confirm that the many statements in the Gospels regarding the Father, Son, and Holy Spirit blend in well with the concept of "Trinity" to describe the nature of God. Just do a word search on phrases like "I and the Father are One", "we will send the Holy Spirit", and "being filled with the Holy Spirit, Jesus ...". The most obvious statement of all of course is the commandment to baptize others in the name of the "Father, Son, and Holy Spirit."

So for those who reject the reality of love, the reality of persons, or the reality of relationships, the existence of the Trinity is a meaningless, a silly theological question like "How many angels can dance on the head of a pin?" For those who do not believe in love, or who do not believe in the value, significance, and wonder of other persons, the personhood of God is an irrelevant question.

However, I would propose that for those who do believe in something called love, who value love, and who value and are amazed at the wonder of our existence, and the existence of other persons, the belief that God is Trinity, not merely a "force", and not merely a single person all alone, makes very good sense out of this amazing story of God and of ourselves.

Here I would like to end with an "aside" (as I mentioned in the introduction) that will perhaps be of more interest to the believer than to those who do not yet believe. It pertains to the appropriate place that the Virgin Mary should have in our Christian faith and spirituality. There are many for whom she has virtually no place at all, and there are those who perhaps felt comfortable with a devotion to Mary that was so extreme that it appeared to be a substitute for devotion to Christ. Both extremes I think are not the best path.

Regarding the latter situation, devotion to Christ has to be the basis of and the reason for devotion to Mary. The inspiration to love and honor the Mother of God should come from the fact that she is the mother of Christ, and because we sense that she can bring us closer to Him. This assumption may often be implicitly there, and so we should avoid being tempted to judge those who to us may appear more devoted to Mary than to Christ, but is an assumption that should never be forgotten.

Regarding the former extreme of giving almost no role to Mary in our faith and in our lives, the topic of this chapter is very relevant. Consider for the moment that the Trinity was not explicitly revealed in the Old Testament. If it were explicitly revealed, one would expect at least some branch of Judaism to affirm the idea of God being Trinity, but to my knowledge that has never been the case. If this is true, then consider the significance of the Annunciation, when the angel Gabriel announced to Mary that she was to become the mother of the Christ, the Son of God. Whether she formulated the concept of "Trinity" or not, she most certainly was the first human being to whom God's nature was disclosed.

The angel said to Mary that, "He will be called the Son of God". The idea of a "Son" has always been in human experience the assumption of a new person who receives and inherits many of the attributes, characteristics, and privileges of the parent. So for an angel to identify this new person as the "Son" of God is quite extraordinary, and quite blasphemous, unless true, in the context of an extremely monotheistic faith and culture such as Judaism. There are other polytheistic traditions where such an announcement would have been no big deal. Right away, Mary was being informed that in today's terms, God is not longer one person, but at least two – a Father and a Son.

But wait, there's more! The angel also said in response to her question of "How can this be since I do not know man?" that, "The Holy Spirit shall come upon you and power of the Most High shall overshadow you so that Holy One who is conceived in you shall be called the Son of the Most High." Assuming that only God can beget God, and since the Holy Spirit seems to be identified as distinct from God the Father (the angel did not say that, "God will overshadow you" or that "the Father will overshadow you") it would appear that the Holy Spirit too is God in a way distinct from Father and Son. The best characterization of the Holy Spirit appears to be words like "Power", i.e. something or someone not visible but which has a very tangible effect. Jesus Himself used the analogy of wind blowing through the trees to describe the activity of the Holy Spirit, which seems like a very apt description.

The Annunciation however was not merely an announcement, and Mary's encounter was more than just a revelation of Father, Son, and Holy Spirit. If it had been merely a revelation, that would have been significant enough. The encounter of Moses with God speaking from the burning bush was clearly a huge event, the story of which is told in many different cultures and religious traditions. However, Mary's encounter with God was far beyond what anyone living previously could possibly have imagined. Not only did God disclose His Triune nature to a human being for the first time in history, He proceeded to envelope this human being in His Triune nature and to become one with this human being in a way that had never occurred before or since.

To say "become one with" may sound extreme, as though I were implying somehow a deification of Mary. I would like to make clear that in no way am I implying that now Mary was elevated to the rank of God in any way, shape, or form. What I am saying is that we must now pause for a moment and take a step back. This Annunciation, this process that occurred can only be compared in human terms to what occurs when a husband and wife come together physically in the act of marriage. This uniting of husband and wife is a miraculous wonder of love, best appreciated perhaps by young lovers for whom marriage is new and amazing. A new human being, a child, may result of this union between husband and wife. The creation of life is clearly one of the miraculous attributes of the husband/wife union.

However, it would devaluate marriage and sexual union if we only regarded it as a mechanism for procreation. Even worse, if we as men only regarded "being married" as a means to having offspring and our wives as merely vessels for that offspring, I think most intelligent people of today would regard that as an incredible dehumanization of marriage and of women. In some cultures, and in certain times of history, this attitude may indeed have been present. In most western cultures, this is no longer the case today.

So a husband and wife, in a good marriage situation, get married because of mutual love and respect, experience a depth of union and sharing that is special and unique to them as a couple, something not shared with the general public. Together, they create new life, and that new life is a tangible sign of the deeper, invisible bond or link that exists between the man and the woman. They are separate, unique, and yet there is a union between them that in some sense allows them to be one.

In the same way, returning now to what occurred at the Annunciation, we have to assume that if we, as imperfect men, do not wish to regard our wives as mere vessels of offspring because that would be a degradation of their true identity, how much more true would that be of a perfect and true God who has chosen this woman to bear His Son? So the next step in the process, once Mary said "yes", was that Mary became the bride of the Holy Spirit. The union she experienced was far more than suddenly discovering she was pregnant. The union she experienced was becoming immersed in the very essence of who God is and in the Love who is God.

Returning now to the Trinity therefore, we must consider the fact that Mary not only was given the revelation of Trinity, and not only became the bearer of God's Son, she experienced a perfect union of love with God the Holy Spirit and therefore didn't just learn about the Trinity, she became united with God as Trinity. She became united to God spiritually by the Holy Spirit and she became united physically in her body in the form of becoming the mother of God's Son. So not only was Mary the first human being in history to become aware that God is Trinity, she is the only human being in history to become fully united to God relationally, spiritually, and physically. Mary is special, and had to be more than just a vessel, because anything less would have been beneath the God who performed this amazing miracle.

It is no surprise then that Mary later said, "All generations shall call me blessed." An interesting statement that may sound rather vain at first. Mary, however, is a wonderful example of true humility as explained to me years ago by the same knowledgeable Dominican I mentioned above. True humility comprises of an accurate knowledge of who and what we are, in comparison with our ultimate model, who is God. We should acknowledge our strengths and weaknesses, our gifts and our faults with equal honesty, because to do less is to show lack of appreciation for the gifts God has given us.

So Mary is significant. She is the ultimate example of what God can do. If He can unite with a human being, and become incarnate in a human physical body, then He can do all things with regards to ourselves and our lives.

Chapter Eleven

Creation

This week I needed to improve a method I had developed in my software for describing the behavior of very rich colored inks used for printing packages. I haven't had a chance to test the new approach that came to me. But a visualization of how to go about it is crystal clear in my head. The experience of creating a better mousetrap than the mousetrap I have built before is rather exciting. Creativity (whether pragmatic and functional like a new device or a new piece of software) or artistic (like a new piece of music or a new portrait) is a unique characteristic of humanity. Cave men lived in caves and bears live in caves, but to my knowledge bears do not leave pictures of their hunting expeditions on the walls.

I think what is cool about creating something worthwhile is that before "creation" something did not exist and after creation it does exist. Creating something in my head is a necessary precursor to creating something externally so that it becomes tangible and visible to others. But creating something in my head is not sufficient.

A wise executive, a founder of Adobe, shared one of his favorite quotes during a presentation: "Anyone who has ever taken a shower has had a brilliant idea." Brilliant ideas are cheap and easy. Reducing a brilliant idea to something tangible always requires a price – the effort involved in creation. The challenge is often to create something that exactly represents the idea. Creating buggy software that doesn't quite work or a picture that is only a preliminary sketch is not too hard – creating software that is actually useful or a picture that someone would want to hang on the wall takes a lot of work.

Now consider the fact that the universe exists. Consider the physical laws that have to operate together perfectly in order for this universe to be. Consider the beauty of the Andromeda galaxy pivoting about its central cluster of stars, held together by gravity and rotating in accordance with Newton's laws. Consider the variety, complexity, and structure of plant and animal life on this planet as contrasted with the barren surface of the moon and other planets.

Consider the fact that the temperature on other planets varies from 90K to 700K and the surface pressure varies from 0 on planets like Mercury to 90 times the pressure of Earth on planets like Venus. By contrast, the sole reason why large quantities of liquid and gaseous water (essential to life as we know it) can exist on Earth is because our average temperature is about 70º F (about 340 K) with an average surface pressure of about 1 atmosphere. At this pressure, liquid water can exist between 273 K and 373 K. If the air pressure were a factor of 10 higher or lower (at our current average temperature), water would either be in primarily a liquid or solid phase or primarily gaseous phase, both of which would be problematic. Likewise, we are fortunate that our range of temperature is about 10% the range of temperature on other planets, at an atmospheric pressure that allows sufficient quantities of both liquid and gaseous water.

Human nature tends to take for granted anything that is extended over time. A rich man takes for granted being rich, because that is all he has ever known. We take for granted the amazing contrast between our lives on this planet and the barren wasteland that exists elsewhere. We take for granted a universe with structure and well-defined laws because that is all we have ever known. I take for granted that my brain is able to ponder these things and write them down in the form of a book, because it has always been this way as far back as my memory goes.

People involved in robotics can appreciate how hard it is to create a fully functional robot. Even with intense effort, there are so many variables that can go wrong that trying to get a robot to perform even a simple task can be maddening. If I simply stopped what I was doing, if I stopped writing the software halfway through, if I only partially constructed the necessary motors, gears, and actuators to make the robot move, how long do you suppose it would take before all the components finally constructed themselves and began operation? If I disassemble a Swiss pocket-watch and shake the pieces up in a bag, how long do you suppose I would have to keep shaking the bag before the pieces re-assemble themselves perfectly back into the Swiss watch?

I do not wish in any way to get into an argument with those who have an expertise in either evolution or in biology. I understand that given enough time and the predictable behavior of organic molecules and amino acids that life as we known it might be expected to come about, and that having come about life will necessarily develop and evolve into more complex systems. I don't wish to argue whether the cards on the table have been dealt in a certain way. The cards are the cards; the hands that have been dealt have been dealt.

I am simply saying that considering the incredible challenge of consciously getting an electronic brain to work correctly in the form of software, or getting a robot to perform the monumental task of picking up a paper cup without crushing it, I believe that the Dealer is not shuffling the deck randomly: He has stacked the deck in favor of life and in favor our existence. I have no intentions of taking Him to task for doing this.

So if I assume that in some infinitesimally small way, the creative activity of our minds is indicative of the activity of the mind of God, God had all creation in all its infinite detail in His mind before He created it. He had each of us in His mind before we came to be.

Considering the concerted effort I must expend to form a logical sequence of operations in software, thereby avoiding random inconsistencies resulting in a "crash" and the infamous "Blue Screen of DEATH," I do not consider it irrational or illogical to presume the existence of a celestial Software Engineer.

"The truth will set you free." For me, there is a great joy and sense of protected freedom in the proposal that we exist within the context of God's Operating System. One of the praised benefits of the UNIX operating system is that although software modules can (if they choose) interact with one another and communicate information with each other, the system ensures that if one module performs an "illegal operation" and crashes, the system as a whole will keep on running, the other modules will still function, and (unlike more primitive systems) it is NOT necessary to reboot the entire system.

In our existence as human beings, individual souls can interact with each other, help or harm one another, but no one can take or destroy your soul except the one who made it. For the Christian, regardless of how messed up we feel as the result of external circumstances in our lives, we are called to trust the Creator of the Celestial OS to keep our souls intact. "He who began a good work within you will bring it to completion."

Chris Edge

Chapter Twelve

Full Disclosure

I have heard the objections from those struggling with questions of faith that they simply cannot accept that God would choose a tiny group of people out of all the people in the world throughout history to whom He would communicate His truths. To do so would be to abandon all the other peoples of the earth and condemn them to ignorance and darkness.

I think it's helpful to affirm that there always have been and always will be good people who hear God's voice, and who have an amazing amount of wisdom and insight into the nature of God by observing what He does in creation and from their human experience. It is for this reason that there is so much commonality and overlap between religions and philosophies, particularly with regard to ethics. We can also rest assured that God's voice speaks to every person, often in the form of our "conscience", which is why to various degrees all of us are to be held accountable for our actions and inactions when we meet our Maker.

However, when it comes to God's "full disclosure" of Himself and of His plan, it should come as no surprise that He chose to do so by involving Himself in the concrete history of a particular nation. And although the salvation of Christ is

73

relevant for all people throughout history, it should come as no surprise that He chose a particular people to prepare for His coming over a 1500 year period, just as he chose a particular woman to become mother of His Son, and decided to exist as a particular human being with a face, a voice, and a real body.

"What our hands have touched and our eyes have seen." I, a human being named Chris Edge, exist as a specific person in a specific period of history in a particular location, culture, and language. Ideas can be broadcast, a person cannot. If all God wanted to do was communicate His values and ideas, a large loudspeaker mounted in the sky would have been sufficient. However, since He wanted not only to communicate Himself, but also to relate with us in a personal fashion, tangible personal interaction within a specific historical context was required.

Just as with our children we prepare them throughout childhood for their life as an adult, God prepared the human race for the arrival of His Son and for the new level of spiritual maturity that would entail for those who followed Him. The 1500-year preparation was also a means of validating that the arrival of Christ was a planned event with permanent significance, as opposed to something that was self-induced or delusional. However, even with the vast number of intricate prophetic references that are contained in the Old Testament regarding the coming of the Messiah, the way He came and the form He took caught many people by surprise.

Chapter Thirteen

An Unexpected Arrival

When we live in the midst of historic times, it doesn't seem all that amazing or romantic. The mundane day-to-day tasks of life seem to encompass all reality.

Performing daily chores in Nazareth seemed just as commonplace then as our daily commitments seem to be now. We expect "miracles and magic shows" (a favorite phrase of mine from Stephan Schwarz's musical "Pippen") from God, and they do not happen, so we assume nothing much is going on.

The most significant and miraculous event in history took place quietly and without fanfare. A young woman became pregnant, a kind-hearted, broken-hearted fiancé pondered how to minimize the damage, and guided only by a dream pretended that everything was fine. Taxes had to be paid. A difficult journey to Bethlehem had to be taken. It was a long painful ride on the back of a donkey, a crude form of transportation for a pregnant lady by today's standards. Prophecy was fulfilled, woman cried out in labor, a chorus of angels sang, and a small barn was the center of the universe for one night.

Life is full of surprises. It is full of pleasant surprises and unpleasant surprises. Very few of us today are doing things or living a life that we would have predicted 20, 30, or 40 years ago. When I drop a rock, what happens next is fairly predictable. We live this thing called existence, and we have no clue what will happen or where we will be in the future. Life is full of surprises. God is full of surprises.

God surprised us in how He chose to enter this world. The countless prophecies prepared the Israelites for a Messiah, but it seemed reasonable to assume that His coming would involve an explicit show of power, as He demonstrated to Pharaoh in Exodus. Nobody saw it coming when He chose to arrive as a helpless, low-income peasant child.

During His time with us, Christ demonstrated that when it was appropriate, and only then, He could and would reshape matter and the forces of nature in a way that could only be considered an act of God, or the act of a God.

An act of God is an event that no human conspiracy could have caused, such as a tornado hitting a building or an earthquake destroying a town. Christ turned water into wine in order to preserve the joy of a wedding celebration. He multiplied a small quantity of bread and fish into a large quantity of bread and fish to avoid starving people fainting on the way home. He calmed a deadly storm and waves to prevent his disciples from drowning. He walked on water. Mainly, He healed the sick and brought the dead back to life. And just as His Father was always the one to take the lead in deciding when, where, and how to perform a miracle, and since there was no other Christ to raise Christ from His own death, the Father raised Christ Himself as proof that He truly was the Son of God.

God surprised us in how He chose to leave this world. In fact, He entered the world only once, but He chose to leave it twice. No one saw it coming that He would allow Himself to be tortured to death. I'm sure that was the last thing His disciples expected, despite many preparatory warnings from Christ Himself. So He left this world a first time, in humiliation and defeat. Then He returned. So now He would of course proceed to destroy the enemies that executed Him and establish His kingdom, correct? Wrong again. He handed over the reigns of his new corporation to His very junior executives, and left us a second time, this time in a much more glorious fashion.

You just can't make this stuff up. We are used to hearing the story, but the story for those who had not yet heard it a thousand times as we have was entirely unexpected. What could possibly have motivated the disciples or their followers to make any of this up? None of them got rich or powerful. Most of them suffered terribly for proclaiming their faith. Peter was crucified, Andrew was crucified, most became martyrs. Yet they stuck to their story, and the story from the multiple sources is pretty consistent, considering they had no internet to trade notes or Microsoft Word to perform cut and paste.

So my scientist brain looks at the totality of the data in the form of hundreds of eye witness accounts spread out over 1500 years leading up to the time of Christ. Trying to keep something like that up for so long after being dispersed and conquered as a people truly seems impossible to me compared to any other human culture I can think of. Then I look at the eyewitness accounts contained in the Gospels and in the multitude of stories that have followed since. I see no evidence of self-gain or self-promotion in this.

The story is a wonderful story. The truth if it is the truth *should* be wonderful and *should* set us free. But the fact that such a complex yet beautifully woven story should be kept alive against all odds by so many people at no apparent benefit to themselves leads me to believe that I can confidently embrace the story as true, not because it makes me feel better or helps me sleep at night, but because the facts of the case all add up to the same conclusion.

In our courts of law, in order to function and in order to make a just decision, juries must be convinced of guilt or innocence beyond a reasonable doubt based on the facts of the case. The jury rarely makes a decision based on something as simple as dropping a rock to prove gravity. The decision has to be made based on human judgment illuminated as effectively as possible by eyewitness accounts, facts, and rational argument.

In the same way, the most important decisions in life such as whom to love, and how to live ones life, have to be made based on our best human judgment. It's not as straightforward as performing statistical analysis of numerical data using a computer model, but then again, there are similarities.

When I look at a graph of a computer model simulating a phenomenon that I have measured, if the similarity between the model and the data is good, and if I haven't "cheated" by creating so many "fudge-factors" that my model can simulate anything, I say to myself "this has to be right."

I have two choices. I can proceed on with accomplishing the hoped-for objectives that I set for myself in performing this project, or I can continue doubting and challenging the validity of the result until the project ceases to be relevant because too much time as passed. My approach is always to do the former. When it comes to research as well as life, my image is that of a sailboat. I need to proceed forward with the best directions I have. If I have to make course corrections in order to reach my destination, that's OK.

From a Christian way of thinking, God can help steer a moving boat to its destination. He cannot do much if the boat sits at harbor – nor will He do much, because He respects our freedom to choose or not to choose. I encourage anyone reading this to take the chance, set out from the harbor, and ask God for the ability to embrace the story that He created because the story, the very Good News, is a wonderful life-changing event. If you sincerely would like to leave the harbor of skepticism but feel you cannot, ask for help. I did years ago, and I have not regretted that request for even a second.

Chapter Fourteen

A Cloud of Witnesses - The Body of Christ

Since I was very young, I felt that I had friends that I had never met. Right up to the end of high school, I felt that I knew Lewis and Tolkien because their books evoked a thrill and a dream of wondrous adventure that exceeded any other books that I read. I did greatly love and appreciate other writers as well – Robert Heinlein, Ray Bradbury, Ursula La Guin, etc. but Lewis and Tolkien both became a deep part of my life. It is no coincidence that they both drew from the same well of inspiration.

Tolkien, whom I perceive (from his letters) to be a stubborn, immovable rock when it came to his opinions and convictions, was a significant influence on his friend. Lewis, whom I perceive to be a man of brilliance and humility, allowed his skeptical atheistic objections to be quietly overruled by the child-like, confident faith and fertile imagination of Tolkien. He went on to be a great spokesperson of the faith they shared – one of the few who could communicate the faith in a way I could receive it.

That was forty years ago. Since then, I have made many new friends that I have never met. Each new friend has a unique gift to share, so it is never dull. They each draw from the same well, but the gardens they grow are as different as a redwood tree from a rose bush.

The inspiration I experience reading about the lives of Christian saints and reading their own words is similar to what I feel learning about the history of physics and hearing first-hand stories about the human beings who made profound discoveries. I am very privileged to have a dad who is one of the last remaining links to a period of history that saw the creation of modern physics.

My dad went to Cambridge in the late '40's and early '50's for both undergraduate and graduate level physics. When he was there, he met Searle who in turn knew Maxwell, the one who formulated the fundamental equations for electricity and magnetism. He was taught by Dirac, who anticipated the existence of the positron and therefore anti-matter. His thesis advisor was Denys Wilkinson who studied under Lawrence Bragg, the discoverer of X-Ray diffraction. Bragg in turn studied under J.J. Thompson (discoverer of the electron) and Earnest Rutherford (discoverer of the nucleus).

Bragg also hired Crick and Watson. My dad would have "tea at the Cavendish lab" with Crick, Watson, and the other grad students where they would talk about how each of their projects was going. Most of the other grad students, including my dad, thought Crick and Watson were nuts for trying to figure out this huge complex molecule using X-Ray diffraction. It would be like trying to untangle a giant ball of string.

When our family would visit Cambridge with my dad, he would point to this window or that window and say, "That's where Rutherford discovered the nucleus," or "That's where Crick and Watson discovered DNA." Becoming part of that legacy was always inspiring to me – now it was inspiring my kids, with the result that my oldest child, Lauren, has finished her Ph.D. in physics.

It is very interesting to me to hear the story of the process by which each of these famous individuals made their amazing discoveries. One thing they all had in common was patience and perseverance, motivated by the thrill of discovery. I think that thrill of discovery is very akin to the experience of those who explored the New World, the inspiration that motivated astronauts to accept the huge risks required in order to walk on the moon, or that motivates climbers to be one of the few who can conquer Mount Everest.

In the same way, I feel privileged to read the lives and words of real people, some of whom have lived in recent times, who are part of the legacy that began 3500 years ago. Understanding how the universe ticks and being part of that tradition is inspiring. Understanding why the universe ticks in the first place and being aware of the meaning and significance of our lives is like seeing a map, and for the first time in our life knowing where we are located and what the rest of the world looks like.

Like the great ones of science, Christian saints showed patience and perseverance in living their faith. Each of us in our own small way must also embrace patience and perseverance in daily prayer in order to discover the invisible reality that they discovered.

The statement of Christian faith, the Creed, speaks of the "communion of saints". Scripture speaks of the "Body of

Christ". Theilhard de Jardin, a Jesuit that I mention below, saw these concepts as an elegant extension of evolution. Consider the pattern: amino acids forming something greater, DNA; single-celled organisms clustering together to form plankton and simple plant life; cells forming complex aquatic invertebrates. These evolve into fish that are vertebrates which evolve into amphibians; then reptiles; finally mammals. Finally, primates evolve in complexity to the point that they are able to be self-aware. It is at this point that God infuses an image of Himself into a primate, thereby creating human beings.

What level of complexity can now be observed after the Incarnation? Theilhard considers that for those who are connected by the Holy Spirit a new phase has begun. Saint Paul's description of a spiritual body comprised of Christ the head and we as parts of the body is a consistent extension of evolution. The cells of our bodies create organs. The organs unite to form human beings. Our selves are greater than the sum of these parts. The body of Christ is a new and wonderful reality that is greater than any one of us individually. This network of members forming this new type of bodily existence spans the entire globe of this planet called earth.

Being part of the body of Christ, the communion of saints, is an experience that is very real and tangible. It may be difficult to imagine for those who have not experienced it.

Being part of a loving family in today's difficult world is a rare gift. It may be difficult to imagine for those who have never felt unconditional love in the context of a family.

The same is true, I think, for those who have never experienced being part of the Christian family. The good news is that we cannot control the family experience we may have had growing up. But we can always choose to become part of Christ's family.

We are taught that we are connected together by the Holy Spirit who is like our central nervous system. I think it is for this reason that I feel that my Christian predecessors are more than historical figures: they are intimate friends.

My own list of friends that I have never met looks something like this:

Henri Nouwen – I heard about Nouwen from a dorm advisor at the University of Virginia. He had taken classes from him at Yale Divinity School. What impressed him about Nouwen was that students would wait patiently in line for an hour after class to confer with him about the deeply personal aspects of their lives. He seemed to have a deep understanding of what they were struggling with and could help make sense out of it, all from the context of his faith in Christ.

Mother Teresa of Calcutta – I heard about Mother Teresa from a close friend during my freshman year in 1974. Her example of love and compassion are now of course well known. Her writings are a humbling reminder that love and service to the poor is an honor, not a burden, and gives more to the giver than to the recipient. It is important to remember that she emphasized that she and her sisters are not "social workers". They honor and love each dying street person they serve with the honor and love they have for Christ.

John of the Cross – I learned about John from my friend mentioned above. John's profound contribution to understanding prayer is the image of the "dark night". This image referred to the fact that when a spiritual person develops in maturity, the good feelings of spiritual elation may vanish, as well as the inspiring insights or images that one may have experienced initially. The correct response is trust and perseverance in prayer, and in loving and serving God and other people, regardless of how one may feel or not feel emotionally.

There are different ways of understanding the dark night.

In the earlier stages, the dark night is part of the process of purification of our souls, and experiencing the Cross in order to be truly "born again" as followers of Christ. Christ Himself experienced this painful stage on our behalf when he said "My God, My God, why have You abandoned Me?"

This stage may pass, and later a new more challenging stage may begin. One way of looking at this latter stage is that God is no longer just an experience or a thought – He is becoming part of who you are as a person. It is reminiscent of the relationship of husband and wife after 30 years of a good marriage – it is beyond mere emotions and sentimentality, but rather it is like being as familiar with someone else as we are familiar with ourselves.

This reality is even deeper and possibly painful in this latter dark night. This is because God has become so united to the person's deepest self, you can no longer "see" Him or experience Him, just as we cannot see ourselves without a mirror. The experience may therefore seem like abandonment. The reality is that the holy person is becoming most like Christ.

No one knew this till after her death, but Mother Teresa apparently suffered this sense of abandonment for many years, yet did not hesitate to continue praying, loving, and serving. The fact that she could do so was testimony to her having become a "little Christ".

Teresa of Avila – another master teacher in prayer. Her helpful image in particular is that our selves, our inner souls, can be regarded as magnificent castles. The goal in prayer is to get past the garbage and the poisonous snakes (i.e. addictions and compulsive behavior) that often mess up our lives, and enter into the chambers of our true selves, at the center of which is the throne on which Christ sits.

Francis of Assisi – there are many anecdotes about his life, as well as his songs and teaching. Suffice it to say that Francis was crazy, joyful, in love with Christ, in love with the beauty of nature, was able to serve the poorest and sickest of his fellow human beings with joy rather than with despair. Like the Virgin Mary, he emulated the truth that God looks with favor on the poor and lowly.

Damian the Leper – was able to serve the outcast lepers on their quarantined island with no fear for himself, knowing that he would eventually join them by contracting their disease.

John Vianney (The Cure of Ars) – would spend all day every day listening to the confessions and struggles of those who came to him, giving miraculous comfort by confronting their secrets that they had shared with no one, including himself.

Queen Margaret of Scotland – a wonderful example of those who are rich and in power serving the poor. One day, she invited a beggar into the castle and gave him the bed that she and the King themselves slept in. The King understandably protested, no doubt irritated at such an irrational gesture, until he noticed for a moment that the filthy beggar had taken on the appearance of Christ.

Catherine of Sienna – illiterate, she taught herself to read and write under the inspiration of the Holy Spirit, served the poor who were dying of horrible diseases, experienced becoming a bride of Christ, and become an influential prophetic advisor to the Pope and Kings of her time.

Maximilian Kobe – gave his life in substitution for a Jewish man in the concentration camp who had a wife and children, as well as wrote deeply on focusing on the love of God which lasts forever rather than the passing pleasures of this world.

Augustine – a brilliant thinker and orator who struggled with his addiction to sensuality and with his confusion and skepticism regarding truth. He finally had the grace to let go and simply believe like a small child. His subsequent insights into the nature of our need for God live on to this day, as captured in this song we often sing in our community:

> *My heart searches restlessly*
> *And finds no rest till it rests in Thee*
> *Oh Seeker, You sought for me*
> *Your love has found me – I am taken by Thee.*

Theilard de Jardin – A Jesuit paleontologist who in no way was threatened by science or evolution, participated in some of Leaky's great discoveries, and in the midst of deepening our scientific understanding of evolution saw the hand of God intimately at work, evolving all Creation towards Christ, the Alpha and the Omega, the beginning and the end.

Iranaeus of Lyons – One of my favorite quotes, "The Glory of God is man fully alive." Irenaeus was a disciple of Polycarp, disciple of John. I appreciate the fact that he writes in a contemporary fashion, clearly has significant first-hand knowledge of the Gospels, and like Theilard, reflected on "recapitulation", the bringing back together of humankind and all creation into Christ the head, who is both creator and redeemer after creation had fallen.

G.K. Chesterton – an overweight, jovial man with a brilliant sense of humor and of humanity. He wrote weekly newspaper editorials for the common man with profound insights into our human frailty, God's goodness, and our amazing universe.

Finally, I simply have to mention **Deacon Lawrence**. The story goes that during the Roman persecution, Lawrence was cooked to death slowly over a grill. Instead of despairing, or screaming out in horror and pain over his plight, Lawrence calmly informed his torturers "I think I'm done on that side – you'd better turn me over."

We are also fortunate to have heroes who are very much alive today. I have not met these individuals personally – someday perhaps, I will have that privilege. Till then, their work and witness fill me hope, and a sense of gratefulness to be part of their family:

Fr. Gregory Boyle - If you only read one book this year, read *Tatoos on the Heart*, a book about Fr. Greg's work giving hope and dignity to L.A. gang members through unconditional love and the opportunity to have a real job for the first time in their lives.

Mary Jo Copland – Go to her website

www.SharingAndCaringHands.org

to learn everything you need to know about her ministry to poor and homeless individuals and families. I recently saw a locally produced video that captured the stories of the many beautiful people whose lives have been blessed and given hope by her praying together with them along with the many services her programs offer – food, clothing, shelter, protection, education, counseling, etc.

Finally – to end my list of today's cloud of witnesses – I have to add my newest hero:

Marsha Linehan – For those who have suffered directly or indirectly from mental illness, depression, anxiety, personality disorders, etc., this name might be familiar. In the course of experiencing mental health issues in our family, we started hearing amazing stories about DBT – Dialectical Behavior Therapy and were encouraged to engage in this program. Later, we heard from a friend that the originator of this approach was a deep Christian woman who based much of her work on the spiritual direction she received from a Redemptorist priest.

Finally and most recently, I read her interview with the New York Times where she shared that she had personal experience in these matters, and was not merely a theorist.

She herself had been hospitalized for two years at the age of 17. Her turning point occurred while praying in Chapel in front of the crucifix. She experienced the reality of God's love and presence in a way she had never experienced before, and this was key to her recovery.

To me this is what true Christianity is all about in the context of this book – not rejecting science and medicine, but rather embracing them. Take all the good human knowledge that exists and combine it with the wisdom and the hope of the Good News in order to create that which is best in class for helping those in need.

The list goes on – each person can and should develop their own unique list of individuals whose stories and writings inspire us in some way and give us hope that our lives are not limited to the joys and sorrows of this world, but rather that this world is only the beginning of a much bigger and more wonderful story.

The list of individuals to choose from is so large that it can blur together as one large community, rather like an entire nation. I therefore appreciate the phrase coined by the author of the letter to the Hebrews – they are a "cloud of witnesses" who are like a multitude of data points all pointing towards the reality of the Son of God.

Chris Edge

Chapter Fifteen

Evidence of Inspiration

It seems to me that God does not attempt to give us proofs of His existence or proofs of the Christian faith. Rather He gives us strong evidence or signs. If a man loves a woman, he does not attempt to draw up mathematical equations to prove his love for her. He expresses his love in a personal way, and gives tangible evidence of his love in the form of a precious ring and in the form of a public commitment of devotion and faithfulness. He gives tangible evidence by sharing all that he has and all that he is with her in his daily existence.

The Christian faith proclaims that Jesus was God's Son. We all know the stories from the Gospels of how He behaved. Would we expect someone who was the Son of God to behave any differently? Would we expect the Son of God to merely pat people on the head and say, "Gee, I'm really sorry you have leprosy and are isolated from all human beings." Or would we expect that, having the power and privilege of God, He would heal them? Likewise, looking upon our human condition of sin and hurt, would He simply say "Gee, that's too bad," and then depart without offering any form of help or assistance, or some means by which our frail, sorrowful existence might be strengthened and made whole? Rather, would He not do everything within His power to make our lives whole, even if it meant tremendous pain and suffering on His part?

If I look at the lives of the countless individuals who have followed Christ in His footsteps, from the time of Christ to our modern times today, I see the ongoing, continued evidence of inspiration. The list of lives inspired to do great things in the way of love, service, and prayer is very long.

Likewise, there is a vast network of mutual validation and confirmation of the truth, much like having a large group of witnesses in a trial whose testimonies all appear to corroborate one another's statements. It is certainly possible, as it is in a trial, that the whole thing is one large conspiracy, where everyone knows that they are lying and have managed to synchronize their lies to ensure they do not contradict one another. But considering the fact that the only reward they would receive for such a cover up during the first 300 years of Christianity was death by torture, what are the odds of that? What would be the motivation?

For example, consider Irenaeus of Lyons. Irenaeus is a well-documented historical figure, partly because he wrote multiple volumes entitled "Against Heresies". The Gnostic cults of his time were really into what we would call today a "secret decoder ring" type of religion. Part of the attraction was the pride in knowing that you had a secret knowledge of "the truth" known only to you and the members of your society.

In order to dispel some of the "mystique" and therefore the attraction of these cults, Irenaeus documented their "secret" mysteries in excruciating detail. In doing so, he made quite obvious how ridiculous they were, even to the point of summarizing 'them in the form of a parody where he renamed all the celestial beings described in their writings with the names of various fruits and vegetables, dominated by the "great celestial watermelon" which explodes in order to disseminate a wealth of watermelon seeds.

To my knowledge, there has never been any questioning by scholars of the authenticity and historical existence of Irenaeus, any more than questioning the existence of Caesar or Socrates.

This is a significant piece of evidence to consider, because Irenaeus was a disciple of Polycarp who in turn was a disciple of John the Evangelist. Neither John nor any of the apostles got rich or had an easy life because of proclaiming that Christ was the Son of God. None of them got anything out of telling the many accounts of His miracles. The same was true of Polycarp and Irenaeus. Apparently what motivated these three men was pure conviction that this was the truth.

Now by the time Irenaeus came along in the second century, all the Gospels and letters of Paul were well known. Irenaeus and his contemporaries make many quotes that indicate familiarity with what we today call the New Testament. If the Gospels or the letters of Paul were in any way fabricated or contradictory to the first hand word of mouth passed on by John to Polycarp and Polycarp to Irenaeus, I'm sure it would have been flagged. Clearly, the story of Jesus coming into the world, proclaiming the Kingdom of God, and performing amazing miracles were consistent with what they had heard over and over from John to Polycarp and Polycarp to Irenaeus.

The story has been given from Irenaeus and his generation to us. The claim is that the story is true and has key implications for how we respond to life and what our life is all about. My response to this declaration is that I have heard the story, the evidence strongly supports that the story is true, and I will therefore embrace the story and live my life in accordance with the message of hope that it brings.

In the same way, I embrace all that I was taught in my physics training as true and perform the task of solving technical problems and creating new opportunities by building on the foundation of everything I have learned. Part of building on that foundation may involve developing a deeper understanding or making better sense out of aspects that were perhaps a bit fuzzy in the past. However, that does NOT mean that I am dispelling everything that I learned as untrue, or that what I learned was bogus. Everything I learned was valid in so far as it goes. Einstein never implied nor intended to imply that Newton was a liar.

We are all free to accept or reject. But before I outright reject the wonderful history of God's love for each of us, let me ask myself the following:

What strong evidence is there that I should disbelieve the story?

Why should I be skeptical about this consistent explanation of what life is all about when I accept without question the very consistent explanation of how the universe ticks from the community of physicists?

Is there an alternative religion or worldview that appears to have a 3500-year history of God communicating his love and mercy to humankind?

Finally, there is the ultimate evidence of inspiration, the culmination of data that can only be acquired through personal experience. I can only truly know that Shakespeare was a great writer in a direct way by reading Shakespeare, not just believing theoretically. I can only know God by reading the words inspired by God and by pursuing Him in my deepest and most inmost self.

Sometimes God gives signs to let us know He is here with us. I have personally known those who appear to have received miraculous healing. One was a mom who attended my first prayer group back in college. She had been diagnosed with terminal lung cancer years before. She was prayed over by friends and the cancer was suddenly gone from the x-rays. By the time I knew her, she had already outlived her doctor.

Likewise, in our faith community, there was a dad twenty years ago who had passed out from carbon monoxide emitted from a malfunctioning gas heater. He was in a coma. The doctors at the hospital sadly told his wife to brace herself that the two outcomes were most likely death or lifetime vegetative state. The wife however calmly said that she would go home to be with their ten children, and that she planned to see her husband when he was awake the next morning. The doctors were sorry to see such denial, because the reality would be so much harder to take when it finally dawned on her.

What the doctors didn't know was that there was a reason why the man was there as soon as he was after succumbing to the CO poisoning. He obviously did not call 911 himself. His wife was attending to their 10 children at home when she felt God speaking to her heart, telling her to go NOW to her husband's green house where he worked each day and tell him that she loved him. She resisted this illogical suggestion – it was a bit of a distance to visit the green house, and she had 10 kids to take care of. But she went – and found her husband unconscious. She was able to get help right away instead of wondering hours later why he had not come home for dinner, and finding her husband dead from CO.

So there was a good reason why this woman responded so calmly to the doctors' dire predictions. And sure enough, next morning she arrived at the hospital to find her husband awake and alert. His first words were "I love you". He was released within a day with no apparent issues. Subsequent testing confirmed the same thing – no issues. There's a great photo that was taken of the two of them, holding a "Get Well" card inscribed with the words "I don't believe in miracles – I depend on them."

When I later told a friend of mine who is a forensic doctor the CO readings they took from the blood samples, he concurred that if he had measured those readings in the blood of a victim that he was sampling for the coroner report he would give a 90% probability that CO was cause of death.

More recently, Randy, one of the leaders in our faith community was on men's retreat. Jim, one of the other leaders, had a dream about him that sounded pretty scary (something like a heart attack) and he asked Randy if he had seen a doctor lately. Randy asked why and Jim just said, "You don't want to know – just go see doctor right away!!" At the same time, two other guys had a similar sense, one of whom is a doctor, after noticing Randy being somewhat out of breath during a walk. Randy got checked out right away, and sure enough, his clogged arteries to the heart were a time bomb ready to explode. The doctors were able to open his arteries and fix him with catheters before he dropped over from a heart attack.

I have never experienced life-saving miracles or dreams quite like the above. However, healing of relationship, love, and forgiveness seem to be consistent themes in the evidence of inspiration that I have personally experienced.

Once I had a dear friend named Victoria who was planning to become a nun. My freshman year, she and I were inseparable as she mentored me in what she had learned about living a Christian spiritual life. However, for various reasons, mainly silly theological disagreements, we had a falling out. Many months went by with little or no interaction. Finally, one night we guys (who by now were all considering priesthood) had visitors at our apartment, and being good hosts, we slept in sleeping bags in the living room so our guests could have beds.

During the night, I dreamed that my friend Victoria woke me up from my sleeping bag and categorically announced, "Good morning Mr. Edge. I am getting married to David Ridgely." Now this was rather bizarre. I had heard this fellow's name as one of many new people who had joined the circle of friends with whom Victoria associated, but absolutely nothing one could consider significant. In my dream, the barrier that had been between us lifted, partly because I could see that she was struggling with this change in direction. So I remember just giving her a big hug and telling her it would all work out. I felt the presence of God in the dream, melting away whatever disagreements we might have had.

It just so happened that the next day, I had a conversation with Victoria on the telephone. The barrier that had formed months earlier was still there, and our conversation seemed littered with awkward pauses where neither of us knew what to say. Finally, to break the ice I said, "By the way, Victoria, I had really weird dream last night." I proceeded to relate to her the strange announcement that I dreamed that she had made. There was a moment of stunned silence followed by "Damn you, Mr. Edge, damn you!" To which I replied, "Gosh Victoria, it was only a dream." She responded by saying, "Well, Sir, I *hope* you are not a *prophet*! Because despite all my efforts to the *contrary*, I find that I am in *love* with David Ridgely!" What followed was a very dear, heart to heart conversation filled with the love and friendship we had known the entire year before.

The following school year, I had neglected to pay in advance for the linen service we students used, and therefore was sleeping on my bunk bed with a sleeping bag the first day of classes. Suddenly, there was a loud "Bam, bam, bam" on the door. In walked Victoria, who primly perched herself on the side of my bed and announced, "Good morning, Mr. Edge. I am getting married to David Ridgely."

Years later, I had another dream, this time about my roommate, John. John and I lived in a house with other students, sponsored by the Episcopal Church next door. John was a nice fellow, but rather awkward. Several of the students who lived in the house were irritated by him – partly due perhaps to a rather "academic" style of talking and relating that just seemed to rub them totally the wrong way.

I never had any real problems with John at all, but neither did I stick up for him when others expressed their not so positive attitudes.

This went on all school year until I had my dream. In the dream, John was sitting there in front of me in a chair. He was staring straight ahead like he was in a catatonic state. His head was bandaged all around. He clearly was suffering from some major head injury. I looked at poor John, and thought of all the times we as a community had been so unkind and let him down. I got down on my knees and begged his forgiveness. I hugged him, and for the first time felt an overwhelming sense of love and compassion for this good hearted person who had difficulty relating to others through no fault of his own. The dream was so vivid that I took care to write it down in my journal. Again, there was a strong sense of God's presence, inspiring me to love this good man who deserved better.

A year went by. I was working for a year near my family, but kept in touch with my friends at the house. It sounded like things had gone from bad to worse in the relationships between John and his housemates. I came back the following year to continue my studies. John was still there, and so were the house tensions.

Meanwhile, John had taken up a new hobby: hang gliding. Every spare weekend, he would drive out to the beautiful Blue Ridge Parkway and enjoy the view as he jumped from various overlooks.

Then one weekend, John was missing. The weather had been bad, not the best time to go hang gliding. Three days went by:

search crews had no luck. Then finally, John was found, unconscious, buried under his glider.

I visited John in the hospital. Sure enough, there he was all bandaged up, unconscious. For many weeks, John was in a coma. Needless to say, everyone in the house, as well as former housemates like myself, had a change in heart. Certain individuals spent many hours with him, day and night, along with family, talking to him, praying with him, expressing their friendship and asking forgiveness.

Finally, John woke up. After a while, he could speak haltingly. As time went on, it appeared that the accident had a big impact on him in many ways, and on his relationship to others.

Next year, John came to my wedding. He updated my wife and myself that he was doing well, although he still had some halting speech and needed a cane to walk. This however was quite extraordinary, considering that the doctors had told him that he would never walk and never function normally. His persistence and faith paid off, and now John was teaching at a boys' school, fully self-reliant.

There was a third dream I had during college that I had forgotten about until 2004. During college, the rocky marriage of my parents finally dissolved into separation and divorce. The recriminations and outbursts of intense anger and emotion were pretty horrible, so the grief and pain for my brother and myself was quite severe.

I remember going to sleep one night in a state of total misery. That night I dreamed that I was in heaven. My mother and father were standing in front of me, side by side, smiling and relaxed. They looked at me and said, "We just wanted you to know that everything is OK now. We've made up and we're good friends." Once again, in the dream I experienced an incredible peace, the presence of God. I woke up with that same sense that even though things were horrible now, someday things would be OK, if not in this life, then in the next.

Fast-forward about 30 years. I was at my mother's deathbed in 2004 at the hospital. She had dramatically lost weight and had collapsed in her home. Finally there was a clear explanation of why she had no appetite and had hardly eaten for weeks. The doctors confirmed that she was in the final phase of stage four cancer, and would probably die very soon. Over the years, there had been very little interaction between my Mom and Dad since their divorce. Dad had consciously avoided interaction because it just seemed to stir things up. However, once he heard the prognosis, he expressed a desire to visit Mom. Surprisingly, Mom was quite happy for him to visit.

What transpired amazed my brother and me. Dad showed up with a small flower – my Mom graciously accepted it. They proceeded to sit down and chat amicably like old friends. There was no denial of what had occurred in their chatting. My mom even stated flat out, "You were a lousy a husband but the world's greatest ex-husband!" My Dad laughed and the two of them went on like this for the next two days till Mom lost consciousness. As I look back, I realize how appropriate my dream was from 30 years before. My Mom and Dad were not yet in heaven, but my Mom was certainly right on the doorstep, preparing to enter.

So it may seem warm and fuzzy, but fullness of life, loving, joyful relationships with wife, children, family, and friends are the data points that I sample directly every day in the course of my normal human existence. These data points confirm the thesis that Jesus is the Son of God and our Savior. There is a serenity knowing that my life matters, is meaningful, and that someone who is bigger than myself compensates for my human limitations.

Seeking reasons for contradicting the message would not be a productive use of my time. I could doubt and question every positive aspect of my life – the love between my wife and myself, my ability to think and do research, or most recently, why should I as a physicist be writing a book on topics that are outside my expertise. However, why waste time questioning the validity of good things when I can simply accept them as good and proceed ahead? Isn't it much better to love my wife, care for my family, pray continually to my God, and serve those whom I have an opportunity to serve? What great blessing does skepticism bring to the picture?

I can personally attest to the overwhelming influence in my own life of following Christ and embracing the joyful truth that has been passed on and shared by the cloud of witnesses. I can attest that God is present in all things, even in the process of capturing data for the purpose of obtaining a Ph.D. in physics. This process for me and for many physics grad students was incredibly arduous.

It required working 24 hours straight to adjust and fine-tune marginal equipment with just enough signal-to-noise to see a signal. Then it entailed *another* 24 hours to capture data desperately before the whole system ceased to function.

Chris Edge

This was repeated week after week. In memory of this technical boot camp, I wrote the following as part of my dedication for the Ph.D. thesis (which by the way had the catchy title "Search for Quasi-Landau Resonance Behavior in the Photodetachment Spectra of the Sulfur and Selenium Negative Ions in a Magnetic Field"):

The Lab

Midnight to morning
Wrestling an unseen foe
Hidden in the underbrush
     of electronic garbage

In blind faith
Confident and fearful
That this dark night
Of soul will end

And that the Master Physicist
Will continue to lead the blind
In ways they do not know.

In conclusion, I am an intelligent guy, but my weaknesses and deficiencies in terms of discipline would have limited my success in physics years ago. I'm a nice guy, but my tendency to hide behind academics required divine intervention. I'm a pretty good guy, but the incredible depth of love and sacrifice required to love my wife and children faithfully and consistently for a lifetime requires a well that is much deeper than the well I seem to have on my own.

I am forever grateful for being granted access to the divine well offered by Christ. It is because of that well that contrary to my nature I am peaceful and confident. No matter what occurs, no matter what external circumstance occurs, no matter what stupid mistake I may do now or in years to come, I am confident that all be well because I am seeking to follow Christ.

It is the nature of sin that most of the time I can only reach a small fraction of the full potential of all I can be and all I can do. It is the nature of redemption (to steal a phrase from a recent author Matthew Kelly) that I can finally be the best and most complete version of who I am and who I was meant to become when I grow up.

It is the nature of sin that well-meaning attempts to do the right thing tend to go awry. It is the nature of redemption that "all things work together for the good of those who love God and are called to His purpose," even my sins and mistakes, once I acknowledge them.

It is the nature of sin that my life can seem so messed up at times, partly from issues stemming from how I grew up and partly from stupid choices. However, it is the nature of redemption that I have been given life, and have it to the fullest.

My hope is that if you can embrace someday the evidence of inspiration that you too will experience your own unique confirmation.

Chris Edge

Chapter Sixteen

Where else can we go?

As a scientist and as a Christian, I feel that I have made a commitment to follow the truth wherever it may lead. It was misguided faith, not mature faith, to condemn Galileo because he stated that the earth goes around the sun. Theilard de Jardin demonstrated that that the story of evolution is the story of God's creative hand at work.

Science can validate that a sunset is occurring. Science cannot capture the human experience that a sunset is beautiful, a masterpiece of art. The scientist can explain why the sunset occurs, thanks to Galileo. The Christian can appreciate the beauty of a sunset and know whom to thank. The scientist who is also a Christian can do both. Why choose either/or? Why not choose both?

The vast network of scientists today take it for granted that we are playing a deadly game of chance with our environment. It isn't that hard to calculate how much CO_2 emissions we put into the air everyday across the planet. The trapping of infrared radiation by CO_2 is pretty basic science. The increase in temperature of oceans and the dramatic changes in the polar regions has been well-documented by satellite imagery. The concerns raised by scientists go back to the 1970s, long before "global warming" became a political football.

As a scientist, I must accept the overwhelming consensus on this topic by those who are experts, even though I have little direct expertise myself.

As a Christian, I must voice my concerns since the impact will be on the entire human family. However, the terrible consequences as so often occurs will primarily impact third world countries and the poor who live in low-lying regions and in regions with barely enough water to survive. They are my brothers and sisters and have a right to sustainable life.

In the same way, I simply cannot reject the message of hope that has been communicated by the vast network of Christians who have lived since the time of Christ, a message that was anticipated by their Hebrew predecessors for 2000 years. The message is coherent and strong. The human evidence is overwhelming. The news it brings is the best news one could hope for or imagine given the world that we all live in.

Having received the message of hope, the next question is how do we respond? What are the options and alternatives?

One response is apathy. It is quite true that to decide nothing is in fact a decision. The message of hope requires a response. If the answer is yes, we receive the great gift of hope, joy, and gratefulness like a small child that has never had parents and is suddenly adopted by a loving mom and dad. If our response is no, we must decide how we do plan to live our lives, and who or what is the basis of our daily living.

What are the alternatives? Science is wonderful, but science only overlaps with the human dimension for those things we can measure and analyze. Science cannot give meaning to the human reality of love, relationship, and personhood – at best science can attempt to explain why they exist, but in doing so reduce love, relationship, and personhood to complicated chemical processes. I refuse to regard individuals as merely bags of organic compounds.

For those who are quite vocal regarding their atheistic persuasion, I would have to ask whether we agree that you (at least) exist. Can we acknowledge that you are a good person, and that your repudiation of faith in God stems from a desire to affirm the truth as you see it? You are unable to affirm the reality of God, and do not wish to be a hypocrite pretending belief in a God whom you believe does not exist.

If we can affirm the above, then *you* are an indication to me of the existence of a good and loving God. Take the key words from the above paragraph – Person, Good, and Truth – take them to their ultimate end there you have the definition of God. He is the ultimate Person (not just a force) who is Goodness and Truth.

Chapter Seventeen

The Cross

I look at all that I have learned about the universe from my training in physics and from my professional life in research and experience great wonder. It is only a thimble of knowledge compared to what many others know and compared to what is out there, but even the little bit that I know gives great joy and satisfaction that our universe can be comprehended. A rock falls consistently if I drop it multiple times. A single scattered electron through a magnetic field may be unpredictable, but a large number of electrons together behave in a predictable fashion. It's like disassembling an elaborate Swiss pocket watch and appreciating the effort to make everything tick with such great precision.

Part of the comfort I guess is the feeling that the universe makes sense, and there is a logical reason for why things are what they are, and why things happen the way they happen. When things behave in a surprising way, we eventually find that it is our limited knowledge and understanding that results in apparent contradiction. Instead of responding with denial of observation, it is always best to respond with humility regarding our limited understanding and seek the deeper truth that is pointed to by the facts.

In the same way, I look at the Cross with great joy and wonder. For the Christian, the Cross is the most complete and succinct summation of God's identity that can be understood by mortal man and woman.

The Cross tells us that God loves us enough to become a man.

The Cross tells us that God is willing to suffer greatly as a man because of His love for us.

The Cross tells us that somehow God brings good out of the evil in this world.

The Cross tells us that God has conquered death and sin for our sake.

The Cross enables us to stare at that which most of us fear most, not just death, but horrible death, and become free and unafraid like a happy, trusting child.

The Cross tells us that somehow God has healed the deepest wounds and sins and dis-functional parts of our lives and has made each of us a member of the Royal Family – the Holy Trinity.

All this and so much more from one simple image, one symbol, a mere two sticks that can be drawn by a two-year-old. What other image or symbol in the history of humanity can I possibly imagine that communicates so much goodness about the nature of God and about our true identity? What more could I possibly want? What else could make our confusing existence more meaningful or significant?

Suffering comes in many different forms in our lives. One of the deepest forms of suffering is that which comes from watching someone else we love suffering great pain, hardship, illness, or persecution. Sometimes all we can do is wait, hope, and pray. Patience is not patience if there is no suffering – patience is the willingness to wait with pain. The message of the Cross is that for those who believe there is always dawn after the darkest night. There is always resurrection after death, and the resurrected life is greater than the life that preceded it. The tree and flower is much greater than the seed that had to die in order for them to exist.

I would like to end with one of the most meaningful poems I have ever read, written by my friend Victoria years ago when I was a freshman in college. I hope it touches you as deeply as it still touches me:

Perfectly poised, and at rest,
Turned inward to the Love that
Indwells to destroy its dwelling
And recreate it to swift life
Once more – Soul, afraid and bold,
Valiant beyond the imaginings
(However wild) of all the heroes
Whose names sweep down to
Even these latter days,
Knows the Source of her testings
And loves the keen of the sword
Even turned against her.

O Love, surgeon of our sickness,
Terrible Prince whose even mercy
Stings and whose love wears alien eyes,
Your goodness is beyond imagining

Chris Edge

The clear light of Your presence
Beyond all suffering. Teach us
In Your being near, the one cross.

www.ingramcontent.com/pod-product-compliance
Lightning Source LLC
Chambersburg PA
CBHW071008040426
42443CB00007B/727